T0317373

Hubris

HUBRIS

The Rise, Fall, and Future of Humanity

Johannes Krause and Thomas Trappe

Translated by Sharon Howe

polity

Originally published in German as *Hybris. Die Reise der Menschheit: Zwischen Aufbruch and Scheitern* © by Ullstein Buchverlage GmbH, Berlin. Published in 2021 by Propyläen Verlag

Polity Press
65 Bridge Street
Cambridge CB2 1UR, UK

Polity Press
111 River Street
Hoboken, NJ 07030, USA

ISBN-13: 978-1-5095-6261-9 – hardback

A catalogue record for this book is available from the British Library.

Library of Congress Control Number: 2024938480

Typeset in 11.5 on 14 Adobe Garamond
by Fakenham Prepress Solutions, Fakenham, Norfolk NR21 8NL
Printed and bound in Great Britain by CPI Group (UK) Ltd, Croydon

For further information on Polity, visit our website:
politybooks.com

Contents

Prologue

This is how second decades begin. We know how the twenties turned out last century; what they will bring this time round remains to be seen. In the first three decades of the twentieth century, humanity was wracked by wars, ideologies, revolutions, economic crises, and, not least, a pandemic. A hundred years on, the prospects aren't much better: at the beginning of the twenty-first century, 9/11 abruptly destroyed the dream harboured in some quarters that we might see an end to geopolitical conflicts, and this was followed by a series of crises, each apparently more dramatic than the last: financial and global economic crisis, years of terror of the Islamic State, global refugee flows and, ultimately, the plight of democracies haunted by self-doubt and by the threat of disintegration. In the late 2010s fear, if not panic, in the face of 'climate collapse' galvanized a whole generation. But soon afterwards the feared annihilation of the very basis of our existence faded into the background when a tiny virus, with no will of its own, let alone a higher purpose, succeeded in paralysing the planet for several years and in bringing to a standstill all but the most basic social activities. What a time to be alive – and what a time to fear for your own life and your life support system! Humanity, it seems, has a genuine hangover – only this one can't be cured by a couple of aspirins.

Climate change, the dawn of the pandemic age, overpopulation, the impending collapse of whole ecosystems, the dangers of global military conflicts: the array of problems facing humanity since the beginning of this new decade is practically endless. But who else can solve them, if not us – this incredible species that can fly helicopters on Mars and even produce oxygen from its atmosphere? A species that manages to feed growing numbers of people, guaranteeing them access to education, clean drinking water, and medical care?

We are, without doubt, the most intelligent being this planet has ever produced. We have come to understand what holds the innermost world together, how it began, and how, together with our sun, it will probably disappear in a giant fireball in a few billion years' time. We think of ourselves as all-knowing and all-powerful, yet we are effectively powerless to escape the self-destructive urge that seems to be hopelessly entrenched in our DNA: a mechanism that positively compels us to expand, consume, and absorb the resources around us to the point of exhaustion.

This genetic blueprint is what enabled us to get where we are today. There's just one problem: that brilliant plan has a small flaw. It doesn't factor in planetary boundaries. Now that we are undeniably approaching those boundaries for the first time after millions of years of evolution, an urgent question arises to which we have yet to find an answer: will our DNA also enable us to live within the available limits, with no further possibility of expansion? Or are we doomed by our genes to keep moving forward until our species runs out of steam?

This book is not one of those that deal with the unstoppable ascent of humankind. But neither is it intended to be one about our inevitable demise. It is the story of an exceptional animal that, thanks to a combination of innumerable coincidences, rose at breakneck speed to the top of the evolutionary tree, eventually conquering every last corner of the planet and harnessing it to its own needs. This unique career began a relatively short time ago, after a whole series of failed attempts. Since the human line split off from the common ancestor we share with chimpanzees and bonobos, countless evolutionary paths have led to a dead end. And only one of them has led to us.

In this book we will look at the first humans and their persistent attempts to colonize the world from Africa. Time and again they were defeated, whether by the climate, by devastating natural disasters, or by other cavemen who held sway in Europe and Asia. We will chart the rapid spread of the modern human – *Homo sapiens* – to America and Australia and the parallel decline not just of other human-like creatures, but of nearly all megafauna of the time. We will see how humans tamed the wolf and how they became their own worst enemy. We will follow our rapacious ancestors all the way to the remotest spot on earth, Easter Island, where their fate foreshadowed what threatens us all today: the destruction of the resources we depend on. Finally, we will turn again to Eurasia, where a long battle determined who the next world rulers would be; and their worst enemies, disease-inducing pathogens, would later prove to be their deadliest fellow travellers and their mightiest weapons, influencing the course of history time and again. This went on until the twenty-first century, when humans eventually came to believe that they had conquered this scourge too – only to be proved wrong once more.

Humans can do anything and should take nothing for granted: such is the message of this book. It was written by the archaeogeneticist Johannes Krause, director of the Max Planck Institute for Evolutionary Anthropology in Leipzig, and the journalist Thomas Trappe. Krause was one of the key scientists involved in decoding the Neandertal genome in 2010 at the laboratory of Nobel Laureate Svante Pääbo. Shortly afterwards he identified a hitherto unknown species of early human from a 70,000-year-old finger bone found in Siberia: the so-called Denisovan, the Asian cousin of the Neandertal. Krause went on to become one of the founders of the discipline known as archaeogenetics, which is constantly uncovering new details of the history of humankind, as well as groundbreaking new insights into it.* And the more pieces of the puzzle are filled in, the clearer it becomes that our evolution, while it may look like an unstoppable rise, has also been beset by constant setbacks.

It is a declared aim of this book's authors to add a few darker touches to the rosy picture many people paint of the past of our own species and, by doing so, to bring to the fore the question of how to turn the twenty-first century into a new chapter of success, not of failure. We don't have the solution either. But we can at least take a closer look at the problem: a problem that is partly rooted in our DNA, and has, not without reason, become part of it. But we are not completely at the mercy of that DNA – or at least we don't have to be. And in this we differ from all other species.

A book about human evolution can narrate and interpret this story from a new angle, but under no circumstances can it claim exclusivity on that score. The story we unfold here relies heavily on the work of international scholars: these are listed in 'Sources' (the bibliography chapter), but as a rule not in the text. We have adopted this mode of

*This is the only point in the book where the authors are mentioned by name. From here on, as a general rule, wherever Johannes Krause's research work is discussed, his involvement is not explicitly mentioned.

presentation simply in the interest of readability; it is not intended to diminish these scholars' contribution to the body of knowledge in any way. Likewise, it goes without saying that the research conducted by the institutes co-directed by Johannes Krause – the Max Planck Institute (MPI) for Human History in Jena up to 2020 and, since then, the MPI for Evolutionary Anthropology in Leipzig (MPI EVA) – benefitted from key contributions from a large number of colleagues without whom this book would not have been possible. The same goes for all those scholars who have provided fundamental insights into human evolution over the past decades – insights that have stood the test of time and can only be reinforced by occasional new genetic data. We stand on their shoulders, too.

We will begin our ride through human history with the oldest decoded genome of a modern human; its DNA was published by a team at the MPI EVA in spring 2021. But before we accompany our ancestors on the incredible journey that began in Africa and led meteorically to the present day, let us peep over the shoulders of the scientists to whom we owe our new knowledge in the first place. What brought us to this summit, from which we now survey the world, not knowing whether we will eventually fall from it? Why were we, and not other great apes, the ones to establish a civilization? These are questions that researchers in the field of archaeogenetics are attempting to answer using a unique approach: not by looking into human brains, but by building miniature new ones. These brains are based on those of an old acquaintance who, a few thousand years ago, came a thankless second in the contest for the crown of creation: the Neandertal.

1

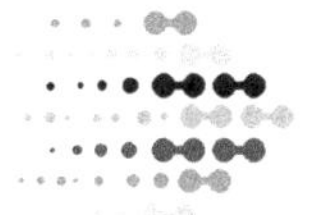

Lab-Grown Humans

A brief excursion into the weird and wonderful world of archaeogenetics: in order to better understand our own brain, we are reconstructing a Neandertal's.[a] And, while we're at it, why not a whole Neandertal, or a *Homo erectus*?

Bring out the Neandertal

One of the places where we are gaining a closer understanding of the extinct Neandertal by bringing parts of him/her back to life is the Max Plank Institute for Evolutionary Anthropology (MPI EVA). This institute is a world leader in genetic research into Neandertals, our closest extinct relatives. In 2010, after years of DNA sequencing and research work, a team led by one of the institute's directors, Svante Pääbo, published the genome of female Neandertals who last walked the earth around 40,000 years ago (all genomes decoded to date are from female specimens). This work won Pääbo the Nobel Prize in Physiology or Medicine in 2022. One of the most important discoveries made at the time was that Neandertals had not really died out, in fact all modern humans outside sub-Saharan Africa still carry genes of these early hominins. Hence early modern humans must have interbred with them when they emerged from Africa to colonize the entire world.

[a] This name comes from a German toponym: the Neandertal valley along the river Düssel. Hence we prefer the vernacular spelling 'Neandertal' to the Latinized 'Neanderthal', which derives from the taxonomic formula *Homo neanderthalensis*. The form 'Neandertal' is increasingly adopted in German.

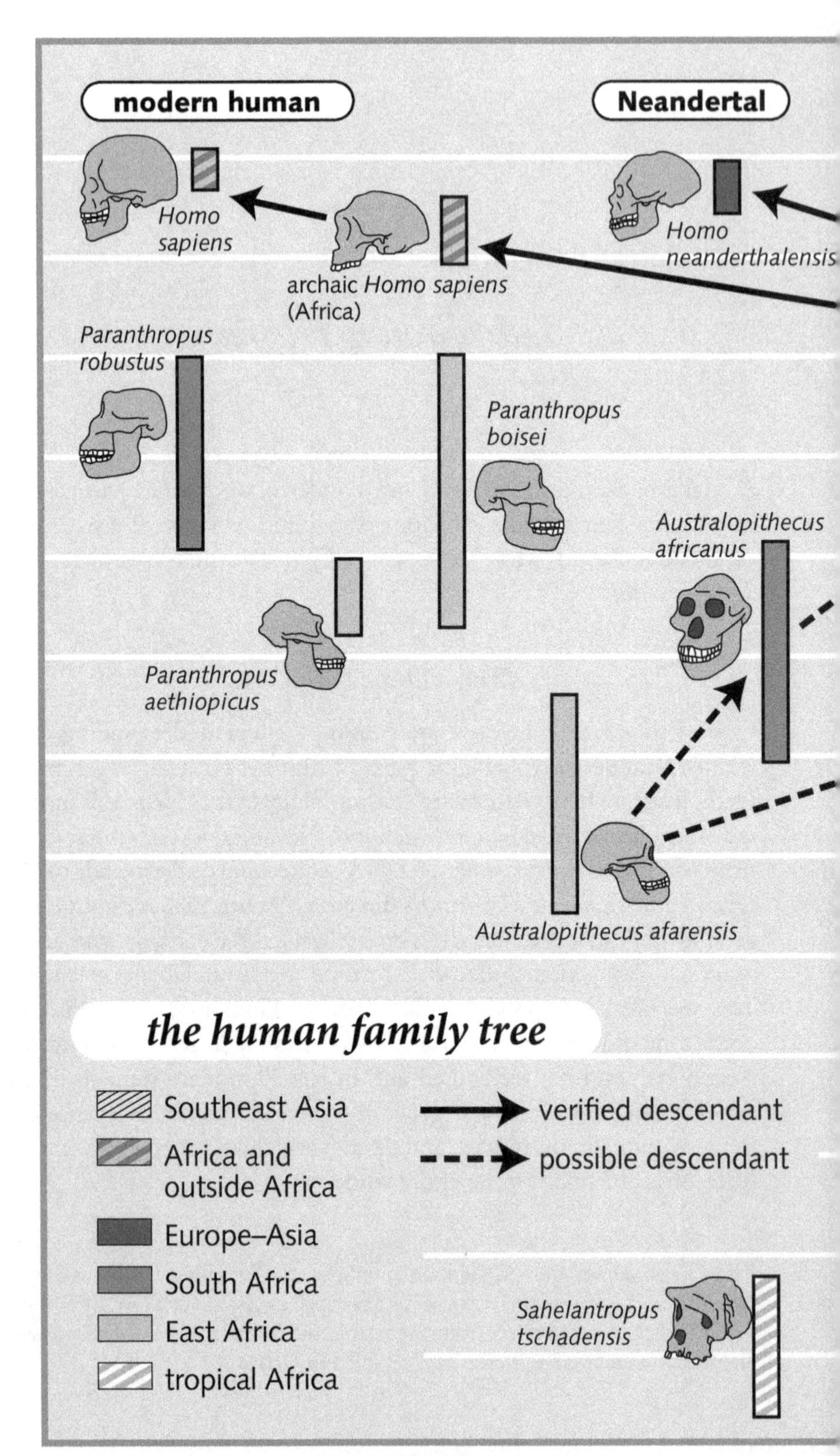

Figure 1 The human family tree.

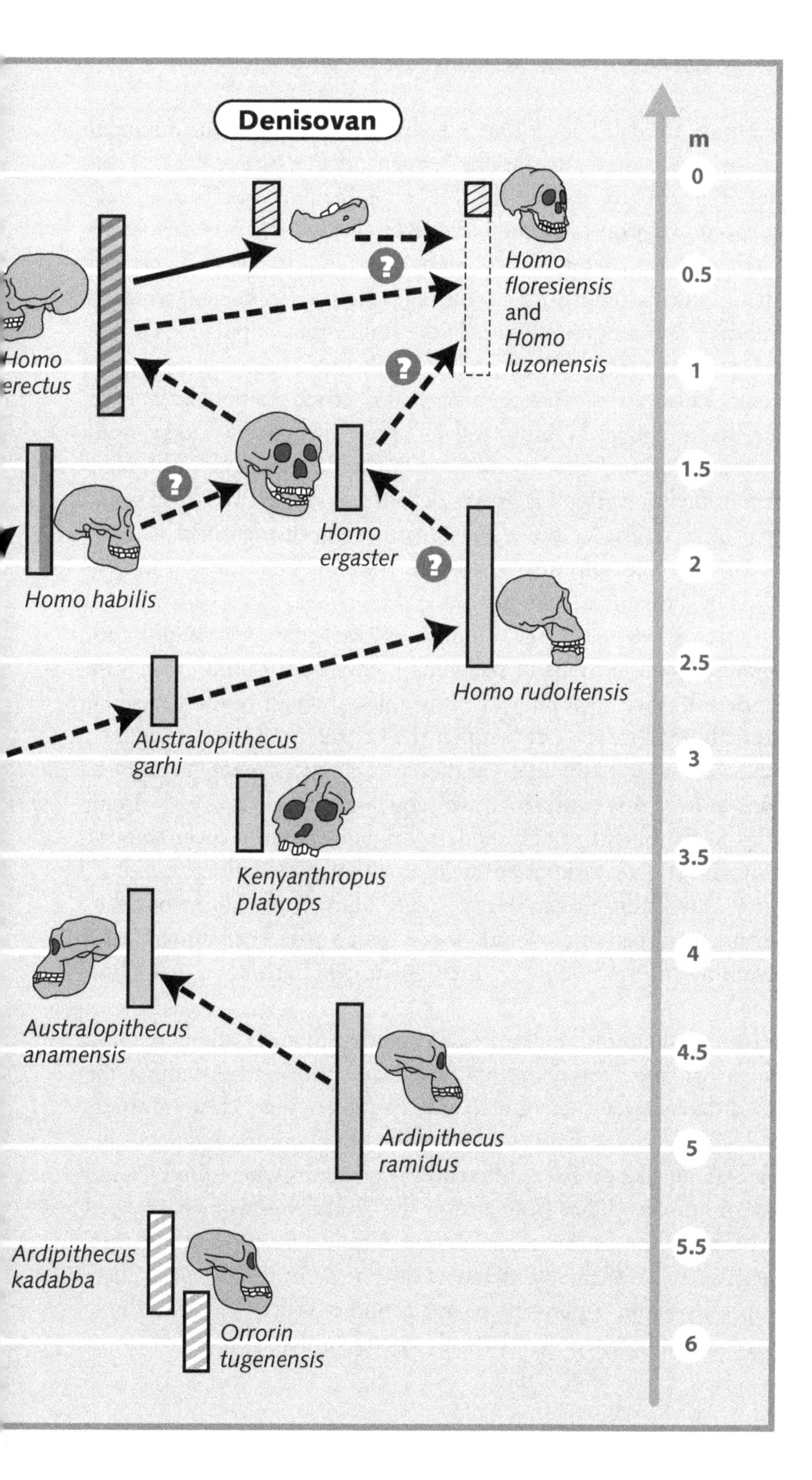

Denisovan
m
0
0.5
1
1.5
2
2.5
3
3.5
4
4.5
5
5.5
6
Homo floresiensis and Homo luzonensis
Homo erectus
Homo ergaster
Homo habilis
Homo rudolfensis
Australopithecus garhi
Kenyanthropus platyops
Australopithecus anamensis
Ardipithecus ramidus
Ardipithecus kadabba
Orrorin tugenensis

Since then, the MPI EVA has consolidated its lead in early human research by proceeding not just to sequence other whole Neandertal genomes, but also to analyse the DNA of Denisovans. This archaic human species split off from the Neandertal lineage at a very early stage and lived in Asia, in some cases alongside Neandertals and modern humans, up until around 50,000 years ago. It, too, left genetic traces in some modern human groups, namely the Indigenous populations of the Philippines, Papua New Guinea, and Australia, which carry an average of 5 per cent Denisovan DNA in their genomes. Crucial to the discovery of this hitherto unknown hominin was an approximately 70,000-year-old finger bone found at the Denisova Cave in the Russian Altai Mountains in southern Siberia whose DNA was decoded at the MPI EVA in 2010. No Denisovan skulls, let alone skeletons, have been identified to date: all we have is DNA from tiny new bone fragments that are periodically unearthed in the same cave in Siberia.

Far more bones – a large number of well-preserved skulls and, occasionally, whole sections of skeletons – have been found in the case of Neandertals: their genome is, next to ours, the best researched of all prehistoric human forms. The fact that the Leipzig team has been able to grow archaic human brain cells – and even miniature organs, 'organoids' – is the result of this comprehensive sequencing work and of a strong similarity to the blueprint of a modern human: the differences amount to a tiny fraction of a thousandth in an otherwise identical genome. Even our nearest non-human relatives, the chimpanzee and the bonobo, differ from us in genome by little more than 1 per cent, although the last common ancestor of these three great apes lived some 7 million years ago.

It wasn't until around 600,000 years ago that modern humans parted company with the Neandertal and Denisovan lineages. Although the genetic differences are marginal, they produce very clear contrasts between Neandertals and modern humans in physiognomy and physique. There are about 30,000 fixed differences – positions where the DNA of all modern humans differs from that of the Neandertal women analysed at the MPI EVA, who resemble chimpanzees at these points in their genome. But most of these differences do not lie in the genes, as these make up only about 2 per cent of our genome. Indeed, there are only ninety genetic differences that actually encode different proteins in the genomes of Neandertals and modern humans and hence are responsible for potentially divergent physical features.

In the past few years, genetic engineering has made it possible to reset a human cell, at certain locations in its genome, to its 'original state' from before the split between modern humans and Neandertals. In other words, it has made it possible to take the genome of a modern human and reverse the evolutionary steps it followed after branching off the line that led to the Neandertals. This is, if you like, a 'neandertalization' of those genomic locations. The process is extremely fiddly and involves introducing into the genetic information of a human cell some of the genetic differences vis-à-vis the Neandertal women. Once this task is completed, the modified cell can grow in a culture into a small clump of brain cells, for example. Such hybrid cells and cell clumps can already be seen in the Leipzig laboratory. The hope is that this would be the next step in the science of evolutionary genetics: we would no longer read DNA differences between archaic and modern humans only from fossilized bones but would observe them directly, in living human cells. This way we would be able to identify the genetic variants that define us as modern humans and are missing from Neandertals. Not all body cells are suitable as a base for neandertalizing human cells.[1] For this operation we need stem cells, which can now be easily produced in the laboratory. At the MPI EVA, this is currently being done using human blood cells, which are then genetically modified with CRISPR/Cas9 'genetic scissors'.[2]

The feasible and the impossible

When it comes to the manipulation of human DNA and the production of hybrid cell structures, the moral implications, though obvious, are by no means entirely predictable, not even for scientists. In 2018, as if to prove that there is also a dark side to our power over our genes, the Chinese researcher He Jiankui, who has since vanished from the academic radar, claimed to have used genetic scissors on human embryos. He justified this molecular biological intervention as an attempt to protect the resulting babies against HIV by modifying one of their genes. He Jiankui never published a paper on his intervention, however. All that the (largely horrified) scientific community got to see was a publicity stunt at an international congress. A year later, the Russian biologist Denis Rebrivok wrote in the journal *Nature* of a plan to edit the genes of human embryos in order to prevent congenital deafness in newborns, albeit with the assurance that he would only do so subject to approval by the relevant authorities. Nothing has been heard of the experiment since.

Cases like these illustrate what a fine line genetic research is currently treading: it is of course easy to conceive of genetic scissors being used to 'neandertalize' a human embryo. In ten years' time at the latest, scientists will have reached the point where they are able to modify numerous genomic locations at once, even without a high-tech laboratory. Unscrupulous researchers wouldn't even need much imagination to achieve a scientific breakthrough of an extremely dubious kind.

At the MPI EVA, the genetic scissors are used to neandertalize human cells, but emphatically not embryos. The aim is not to breed Neandertals or archaic humans – or even whole organs – but merely cell clusters (see

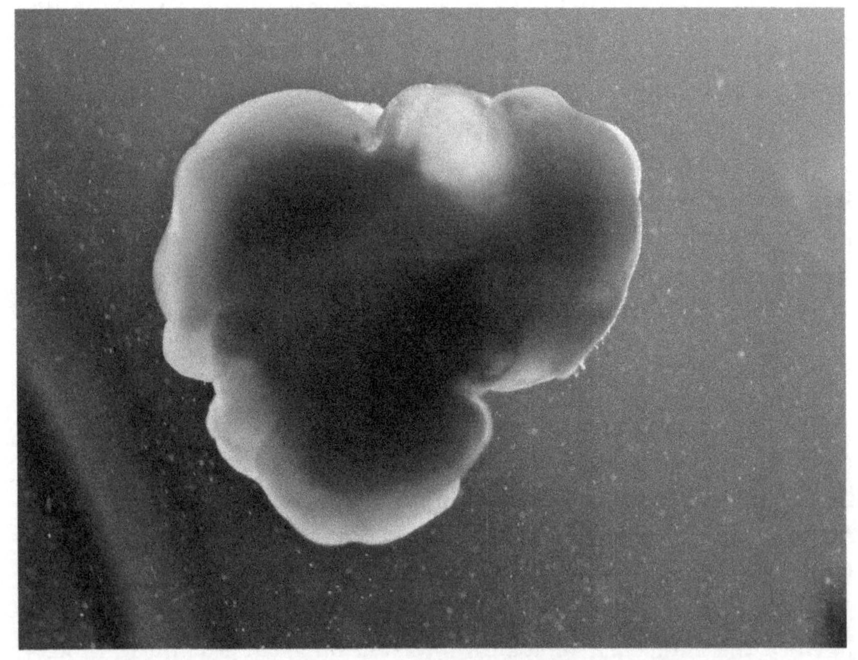

Figure 2 A cell culture 'bred' from brain cells © Daniel Wolny
This already produces biochemical processes that can be observed in the laboratory, although such cell clusters are still a far cry from real organs.

Figure 2). For these, too, can be used to observe biological processes such as the contractions of a heart muscle or the growth and interactions of brain cells.

So far, eight genetic differences between humans and Neandertals have been introduced into the cell cultures grown at the MPI EVA. But it will be a few more years before a cell culture can be grown with all ninety genetic variants. That said, the exponential acceleration we have seen since the turn of the millennium in the field of genetics and, by extension, archaeogenetics is likely to continue. By the end of the twenties it should thus be perfectly possible to incorporate into a human cell not just the ninety genetic differences that separate us from the Neandertals but all 30,000 genetic locations where all humans differ from the Neandertal genome. That would include those bases in the genome that don't encode any proteins but may still fulfil a function.[3]

The Frankenstein genome

For the record, the ninety genetic differences are not the only differences between humans and Neandertals, but they are the only ones between *all* humans and *all* Neandertals. In other words, none of the million decoded genomes of modern-day humans looks like that of a Neandertal in any of those ninety locations. This means that they cannot have developed anywhere but in modern humans, and must have asserted themselves when our ancestors interbred a second time with the Neandertals. So clearly these variants, or at least some of them, must be integral to being human. Nevertheless, there are other segments of the female Neandertal genome that we still carry to this day: all humans outside Africa have an average of 2 per cent Neandertal DNA.[4] In some people, the Neandertal genes are responsible for a particular skin texture, in others for an immune response, and in others for nothing at all – or at least nothing that we can identify.

When the successful decoding of the Neandertal genome was announced in Leipzig in 2010, it was based on a kind of Frankenstein genome, concocted from a mixture of female Neandertal specimens found in a cave in Croatia. What was justifiably celebrated as a big breakthrough back then would hardly make it into a major scientific journal today, given the crude database, which effectively involved throwing into a pot the DNA extracted from three Neandertals and looking to see how far it differed from the modern human genome, decoded barely ten years earlier.[5]

Nevertheless, that was enough for the anthropologically groundbreaking discovery that all modern humans outside Africa carry Neandertal genes, meaning that our ancestors had sex – and evidently plenty of it – with this early hominin. The Leipzig team was only at the beginning of its journey, however, and could do no more than shine a spotlight on an era that could hold the key to a deeper understanding of our own species. By this we mean an understanding of why it was modern humans – and not Neandertals or Denisovans – who went on to colonize the entire world and subjugate or destroy all other life on Earth.

In the meantime, dozens of Neandertal genomes have been decoded, all of them at the MPI EVA. New publications are constantly appearing that discuss evidence not just that Neandertals mated with modern humans, but that modern humans mated with Denisovans and Denisovans with Neandertals. Over the past few years, there have been some impressive discoveries regarding both of these early human species; but such discoveries can only hint at the secrets that lurk in the genome. We may have the blueprint of the Neandertal, with all its bases, but it is still only a blueprint – not a living, breathing individual. To gain an even better understanding of the Neandertal, we would need a real-life one.

Humans are predictable

So far, the idea of bringing our extinct relatives back to life is no more than idle speculation, though this doesn't rule out the possibility of such experiments in future. Still, such a venture wouldn't be an advance in stem cell research but rather a perversion of it. Although the cell clusters cultivated at the MPI EVA are a kind of minuscule forerunner of organs with Neandertal DNA, they wouldn't be able to develop into anything like a whole organ, let alone become part of an organism through transplantation.[6] Nevertheless, such cell cultures are incredibly valuable for archaeogenetic research and could lead to the next big breakthrough in Neandertal research – one whereby cellular processes in these early humans no longer have to be theoretically deduced, but can be directly observed.

Ten for the price of one

As in the early days of archaeogenetics, it was thanks to relatively small technical innovations that scientists in the 2010s were able to delve ever deeper into early human DNA, and hence into the past. Most of the progress achieved in archaeogenetics depends on apparently very simple tricks. Thus the team at the MPI EVA developed a technique that allows for information to be obtained not just from the DNA double helix but also from individual single strands that have survived over the millennia. This way the extracted quantity of DNA can be increased tenfold, so that we can now sequence even the most ancient bones, from which most of the genetic information has already been lost. This is how researchers managed, for example, to sequence the oldest decoded Neandertals to date, who lived in the Sima de los Huesos (Pit of Bones) cave in Spain some 420,000 years ago. And the same technique allowed the reconstruction of a high-quality genome that belonged to a roughly twelve-year-old Denisovan girl from something as tiny as a 70,000-year-old finger bone.

Another achievement of archaeogenetics in recent years was the ability to calculate lines of descent. By comparing early human DNA with the genetic material of modern humans, we are gaining an increasing understanding of which human lineages diverged or split off from each other and when. We simply need to look at the number of genetic mutations in a sequenced genome using a 'gene clock' or 'molecular clock'. The more of these genetic differences we find, the earlier the split occurred.[7] Thus the genetic differences between chimpanzees and all hominins suggest that our last common ancestor lived around 7 million years ago. The common forefathers of modern humans, Neandertals and Denisovans, lived roughly 600,000 years ago; Neandertals and Denisovans parted company some 500,000 years ago. Similarly, we would probably never have been able to study the interbreeding between the different hominins by using the techniques of classical palaeoanthropology alone. Thanks to archaeogenetics, we can now track it in the genome of every human being.

How efficient is a Neandertal heart by comparison with a human one? What metabolic processes can a human liver perform that a Neandertal one can't? Can a Neandertal tolerate alcohol? Or – and this is of course the really big question – are our brain processes different from those of this extinct hominin? Do we form neural networks more quickly, for example? The assumption underlying these questions is obvious: namely that, at some point after the split between humans and Neandertals, changes occurred in our ancestors' brains that led us to turn the world into the place it is today. What is clear is that sheer size is not a factor: the average Neandertal brain weighs 250 grams more than that of a modern human.

That the idea of resurrecting Neandertals for research purposes is not purely theoretical and is undoubtedly at least discussed in many a laboratory (though probably over a second glass of wine rather than the first cup of coffee) was proved a few years ago by George Church. A pioneer of DNA sequencing, Church was a key contributor to the Human Genome Project and in 2006 launched the Personal Genome Project, whose aim was to sequence the genomes of as many individuals as possible, for medical research purposes. In short, he is regarded as an authority among geneticists.

For this reason alone, it is worth considering the other big idea that Church outlined (among other things) in his 2012 book *Regenesis*: that of breeding Neandertals. The main foundation had already been laid, he claimed at the time, with the sequencing of the Neandertal genome. A possible next step would be to break down the genome into thousands of components in order to gradually transfer more and more Neandertal genes into a human stem cell line. The end result, according to Church, would be a 'Neandertal clone', although he was careful to state that such a procedure would require a society-wide debate. The benefit, he argued, was at any rate clear: a greater 'diversity' of the human community that would be conducive to the survival of all species, including humans.

Church did not assume that modern humans were necessarily more intelligent than Neandertals: indeed, the latter's larger brains could equally well indicate the opposite. His point – as he explained in an interview with *Der Spiegel* – was that, if humankind one day had to 'deal with an epidemic, quit the planet, or whatever', the Neandertal 'mentality' might be 'an advantage'. So then: had Neandertals not drawn the short straw in the evolutionary contest, would they have become the better scientists? Instead of contenting themselves with reconstructing the cells of extinct hominins, would they have gone on to conquer global epidemics, antibiotic resistance, and climate change? Or would they have avoided the path that led to all these problems in the first place? These would be questions best put to the Neandertals themselves, were we to follow through Church's vision.

Even today, cloning a Neandertal is still a matter of science fiction, and there is little to suggest that the situation will change (see Figure 3). As a Plan B, Church floated the idea of producing a hybrid by introducing into the human genome specific mutations that distinguish Neandertals from modern humans. The advantage of this method is that it would be selective in terms of the implanted genes, allowing us to cherry-pick the most useful characteristics of the Neandertal. According to Church, the same procedure would apply not just to Neandertals but to any other hominin whose genome was decoded. This way we could travel up to a million years back in time, theoretically bringing the humans of that era – or at least parts of their DNA – back to life: a shopping tour, you might say, in the department store of human evolution.

From an archaeogenetic perspective, however, such a projection is a bold one to say the least. The fact that we have now managed to sequence

Figure 3 Neandertal © Tom Björklund

Also thanks to archaeogenetics, we now have a fairly accurate idea of what Neandertals looked like, although not much is known so far of their social behaviour. These hominins probably lived in close family units too – much like us.

almost all the Neandertal genome is due to a rich fund of well-preserved bone finds from which the DNA could be extracted, the oldest sample being just over 420,000 years old. And, even though countless remains of *Homo erectus* – the most likely common ancestor to us, to Neandertals, and to Denisovans – have been found all over the world, so far they have been only of anthropological, not archaeogenetic, use. If we wanted to reconstruct the genome using the same method as the one used for Neandertals, we would need not just any old bones, but ones from which DNA can be extracted – something that is not available to date.

A much more realistic option, and one that actually takes Church's ideas from 2012 a step further, would be to calculate the genome of *Homo erectus* on a computer. This could be done by considering three genomes: that of a chimpanzee, that of a modern human, and that of a Neandertal – all of which are known to derive from a common ancestor. On the basis of the differences between the great ape on one hand and the two hominins on the other, we could first determine which specific mutations occurred after the common lineage of modern humans branched off from Neandertals. In the gene positions where we are identical with chimpanzees but not with Neandertals, only the Neandertal DNA will have changed – and the same applies the other way round, that is, in gene positions that differ only in modern humans from their counterparts in chimpanzees.

Given this, all positions in the genome of a modern human could then be restored to the 'original state' using gene scissors. The result, however, would not be a pure *Homo erectus* as it existed in Africa a million years ago, but a *Homo sapiens/erectus* hybrid in which the genes of modern humans have been reset. From a medical ethics perspective, the very idea of this is outrageous, but this is precisely why it cannot be brushed aside. Nowadays such a hybrid genome could be calculated on a standard Notebook.

Because we can

The re-creation of a Neandertal in Church's laboratory – or at the MPI EVA, for that matter – is certainly not on the cards, either now or in future. But apart from all the ethical issues, there is also a much more mundane argument against such an experiment. For one thing, if we really wanted to resurrect an archaic species of humans at population level, we would have to create not just one, but hundreds of Neandertals. Since there could be no justification for excluding them from modern human society, or indeed for locking them up, they would sooner or later begin – as they did 50,000 years ago – to interact with us sexually (and we with them), and this would result in hybrid children. Within a few generations, probably in less than a hundred years, the gene pool of this tiny number of Neandertals would sink into that of the billions of modern humans, before presumably disappearing without a trace – apart from the 2 per cent of Neandertal genes that people outside Africa already carry (see Figure 4).

In the end, we would probably be looking at decades of ethical debate and runaway costs, without the benefit of any real groundbreaking discoveries – apart from the fact that the modern human is positively incapable of refraining from doing something just because s/he can. But this hardly needed proving.

In our quest to solve the great riddle of *Homo sapiens*, the neandertalized cell cultures in our laboratories can be only a tool that, with a bit of luck, we can use to identify the single genetic change that gave modern humans their decisive advantage. Was it our culture-building capacity that led to complex, collaborative societies and enabled an increasing specialization of the individual for the benefit of the community? Or was it rather our cruelty towards fellow humans, and even more towards those who didn't belong? The willingness to risk our own life in order to enhance it in unimagined ways? Was it all just a weird coincidence that modern humans ended up on the right evolutionary path, while Neandertals and Denisovans did not? Or is it ultimately the wrong track, a dead end towards which we are currently heading at full speed? And what is it in us that makes us think it would be a good idea to have a Neandertal clone in the passenger seat on that final spurt?

So, for now, let us consider the genetic reconstruction work with neandertalized cells that is currently going on at the MPI EVA as a valuable technological tool that may one day help us to answer these questions. And, in the meantime, let us take a journey far back in history, in the spirit of classic archaeogenetics: a journey to a time when large areas of the northern hemisphere were covered in ice and this part of the world was dominated by Neandertals and Denisovans. And somewhere

Figure 4 The great journey north from Africa © Landesamt für Denkmalpflege und Archäologie Sachsen-Anhalt / Karol Schauer

When modern humans first set out on their great journey north from Africa, they encountered the mostly bitter cold mammoth steppe – a vast hunting ground that laid the foundations for our ancestors' subsequent expansion.

in the Bohemian Forest, not far from present-day Prague, a woman was laid to rest – a woman who, according to prevailing archaeological doctrine, shouldn't have been there at all, and who was an early harbinger of the Neandertals' and the Denisovans' demise. It was this woman – Zlatý kůň – who yielded the oldest genome of a modern human ever to be sequenced.

2

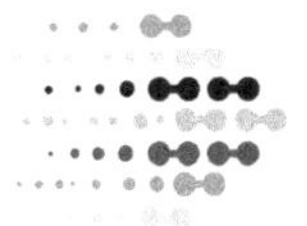

Famine

We enter the bitter cold Ice Age. Our ancestors didn't stand a chance in the north: here it was their cousins, with their rough table manners, who held sway. We meet Zlatý kůň. Our ancestors find a way of coping with the depressing notion of their own mortality. In the background, hyenas prowl.

Dining with one's own kind

Other than in caves, the Europe of the Ice Age (see Figure 5), which only ended around 11,500 years ago, was barely habitable without a good fire. And this applies to all human species who lived here. Caves were the centre of human life at that time, and it is no coincidence that they, along with graves, are the most fruitful sites for archaeologists. The finger bone that opened our eyes to the Denisovans came from Denisova Cave; Chauvet Cave in France contains impressive portraits of aurochs and horses thought to have been painted by our ancestors over 32,000 years ago; and the world's oldest flutes, made from bird and mammoth bones, were found in Geißenklösterle Cave in the Swabian Jura.

From El Sidrón Cave in northern Spain, on the other hand, we have a remarkable relic of food culture, in this case from a group of Neandertals from roughly 49,000 years ago. These individuals appear to have devoured huge quantities of a very specific meat: Neandertal flesh – from children and the elderly, men and women alike. All this was found by a team of archaeologists in a pile of some 2,000 bone fragments deposited in a space no larger than five square metres and nibbled clean. This cave discovery is just one piece of evidence among many to tell us that, for

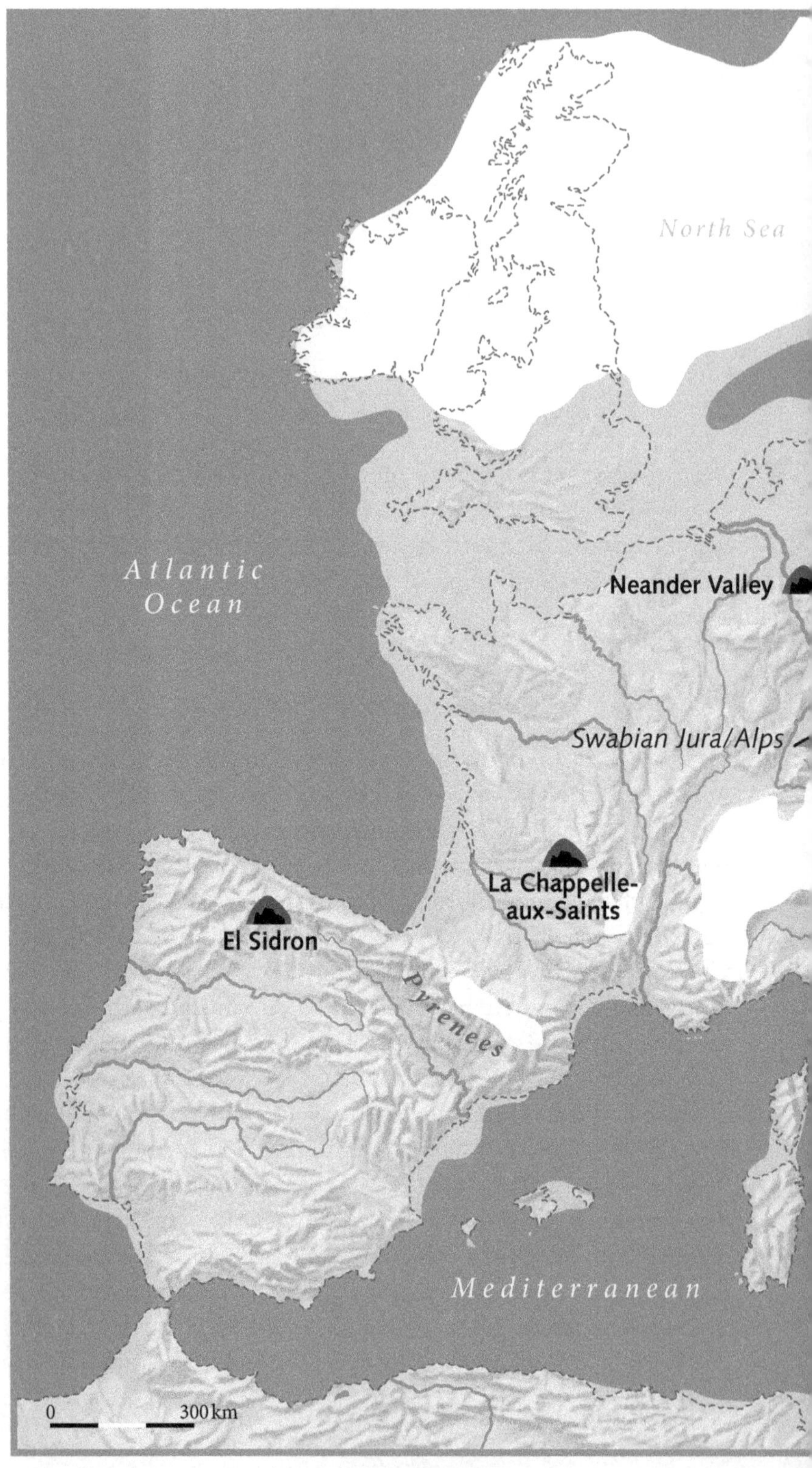

Figure 5 Map of Europe during the last Ice Age.

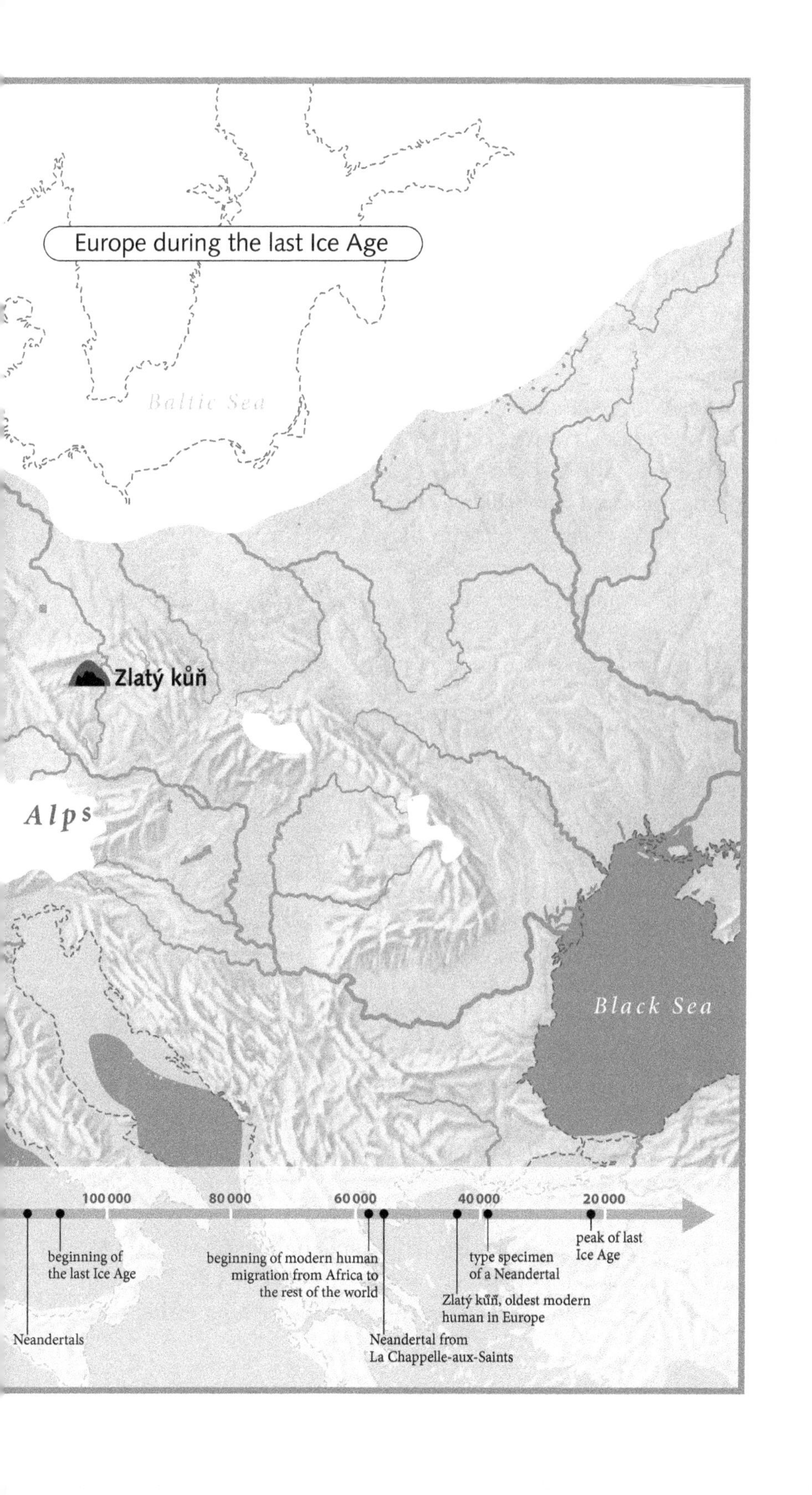
Europe during the last Ice Age
Baltic Sea
Zlatý kůň
Alps
Black Sea
100 000
80 000
60 000
40 000
20 000
beginning of
the last Ice Age
beginning of modern human
migration from Africa to
the rest of the world
type specimen
of a Neandertal
peak of last
Ice Age
Zlatý kůň, oldest modern
human in Europe
Neandertals
Neandertal from
La Chappelle-aux-Saints

whatever reason, Neandertals ate their own kind. For archaeologists today, Neandertal bones whose owners were wholly or at least partially eaten are the rule rather than the exception – a phenomenon not found in the early modern humans who lived in Europe during the Ice Age and of whom Zlatý kůň was an early representative. By all appearances, our ancestors preferred to bury their dead. There was, then, a fundamental difference between Neandertals and humans in their respective dining and burial customs.

That the bones found in El Sidrón Cave came from previously consumed Neandertals can still be detected today, even without forensics. At least thirteen individuals in all age groups, including babies, were found in that pile and bore striking cut marks – some of them on a part of the skull that would make sense only if the victims' tongues were cut out. Hands and feet were removed from the limbs, quite clearly in order to make it possible to nibble away the flesh between the bones, just as one would do with grilled chicken drumsticks today. As for the arm bones and long bones, that is, the thigh and the shin, these were all smashed open for the bone marrow to be sucked out. All the evidence from El Sidrón paints a gruesome picture from our modern perspective. Both the sheer mass of victims and the methods – which look anything but holy – of dismembering and dissecting their mortal remains argue against any kind of ritual killing. These humans are unlikely to have been sacrificed to a god or some other higher being: they were slaughtered like animals and then consumed accordingly.

Although Neandertals were actually keen big-game hunters, there still seem to have been periodic cannibalistic attacks by one group on another. And there are quite a few other records, and very similar archaeological sites in other parts of Spain and in places like France and Croatia. It was probably not a culinary preference that prompted these massacres, but rather dire necessity, as indicated not least by the marks on the victims' bodies. In El Sidrón there are signs – on the teeth found in the bone heaps and on the bones themselves – that all point to the kind of growth disorders seen during persistent famines in the Middle Ages, for instance.

It is therefore quite possible that the Neandertals acted out of sheer desperation, like the rugby players of the Uruguayan Old Christian Club, who survived the legendary plane crash of 1972 by eating their dead teammates after being trapped for days in the ice of the Andes. And it may be that the primal humans from El Sidrón were just as traumatized by their actions as those sportsmen who, after being rescued, spoke of an emotional horror that would haunt them for the rest of their lives. It is equally conceivable, on the other hand, that the Neandertals didn't eat dead members of their own group, but simply chanced upon a large, already weakened group while they were out hunting. The fact that many Neandertal bone finds show clear signs of combat suggests that the groups may indeed have preyed on each other, although many of these injuries are likely to have been sustained while hunting the big game that formed the bulk of these primal humans' diet. The structures found in their skeletal remains find parallels today only in one professional group: rodeo riders, for whom broken bones are par for the course.

Archaeological finds such as those of El Sidrón conform to the image that many people today have of Neandertals: we see our relatives predominantly as rough and brutish, incapable of empathy or anything approaching a culture of their own (see Figure 6); creatures who, when they ate their own kind, did so purely out of hunger and not from any underlying philosophy – like some people in certain parts of the world who, right into the twentieth century, used to consume the brains or hearts of their enemies in order to absorb their strength and energy. Even among Neandertal researchers there are differing schools of thought. One tends towards the widespread notion of the culturally inferior Neandertal, incapable of higher thoughts, while the other is more inclined to believe that our distant relatives were, in every respect, much closer to us than hitherto assumed. And there is indeed some evidence that they were capable of empathy and had something like a sense of family.

For example, the type specimen from the Neandertal Valley – that is, the individual discovered in 1856 after whom this prehistoric human type is named – had a severely crippled left arm broken in many places long before his death. This would have made it impossible for him to hunt alone, meaning that he must have been looked after by other members of his group. One Neandertal found in Croatia had evidently had his whole arm torn off or bitten off by an animal during his lifetime: in a cruel, egotistical Neandertal society this individual wouldn't have stood a chance, yet his tribe evidently granted him one.

Figure 6 An artist's impression of a Neandertal woman © Tom Björklund

An artist's impression of a Neandertal woman. Just like our African ancestors, but also like the early European hunter-gatherers, these archaic humans are thought to have been dark-skinned.

The same goes for the Neandertal male from the French site of La Chapelle-aux-Saints, who had lost all his teeth and appears to have had his food specially prepared for him.

Savaged by hyenas

The woman whose mortal remains were found near Prague in the 1950s fared little better than the Neandertals of El Sidrón Cave, at least according to the marks on her bones, specifically on her skull. These show clear signs of a brutal death: she appears to have been torn to pieces by a cave hyena, an animal indigenous to the whole of Europe and Asia at that time. In 2021, nearly seventy years after the discovery of her skull, we learnt the story behind the find. As a result, the existing narrative of the migration of modern humans from Africa to Europe has had to undergo a major correction.

The woman in question is named after the hill above the cave system in the Czech Republic where she was found: Zlatý kůň, meaning 'golden horse'. She lived in Europe at a time when, according to the hitherto prevailing orthodoxy, no humans other than Neandertals were supposed to exist there. While until 2021 archaeologists assumed that they were dealing with a roughly 12,000-year-old skull, we now know, thanks to genetic analysis, that this one was much older: the woman probably lived 47,000 years ago. What's more, Zlatý kůň's ancestry turned out to have a close affinity with the common ancestors of all people from outside Africa today; so she was relatively close in time to the common tribe that split off in Africa about 80–70,000 years ago. A dash of Neandertal DNA got mixed into this common lineage some 50,000 years ago, and Zlatý kůň's line must have emerged just a few thousand years later (see Figure 7).

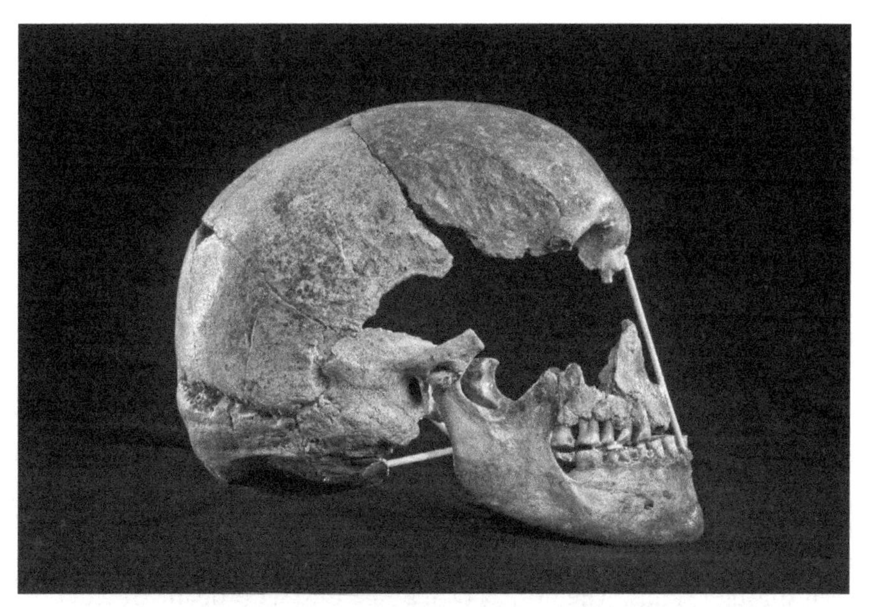

Figure 7 Side view of the virtually intact skull of Zlatý kůň © Martin Frouz, Anthropological Department of the National Museum in Prague

DNA analyses performed in 2021 revealed its owner to be the oldest known modern human from Europe.

Given that the lineages leading to present-day Europeans and Asians split less than 45,000 years ago, this means that the roughly 2,000 years older Zlatý kůň did not produce any descendants still in evidence today. Consequently her genetic trail disappeared – perhaps partly into the stomachs of cave hyenas, which were prolific at the time and which, as the name suggests, occupied the same habitat as modern humans and Neandertals in Ice Age Europe and Asia.

The remains of the prehistoric European woman were found by miners in what was then Czechoslovakia. The limestone cave near the capital city of Prague is part of the country's largest contiguous cave system. In one of the biggest caves below the hill is a 50-metre-high stalagmite, and it was at the foot of this rock formation that the well-preserved skull was found, together with a long bone, nibbled at by hyenas. Alongside these were also a couple of stone tools that, like all stone tools of the period, are recognizable for what they are only by archaeologists, being little more than sharp-edged stone flakes that the untrained eye cannot tell apart from ordinary stones.

There is much evidence to suggest that the tools were not used by people in the cave but subsequently washed up there. With a little imagination, and ignoring signs to the contrary, you could also picture this place as a sacrificial site: the mighty stalagmite would certainly provide a fitting backdrop. But this is an unlikely scenario, as wolf and hyena bones were also found in the cave: these animals must have lived there at the same or at different times, bringing their prey back to the cave to devour it. While they mostly preyed on herbivores, no doubt they also took humans who were weaker or got separated from their group. The Zlatý kůň woman was probably found already dead and dragged into the cave by hyenas, who eat carrion as well as fresh meat. Either that, or she was mauled alive.

Anyone who has ever come face to face with a hyena can imagine how ineffectual a stone or a stick would be against these creatures, which are essentially a set of jaws on legs and can easily tear through bones – or even through skulls. The hyenas of Ice Age Eurasia probably reacted to the first human immigrants as a guard dog does to a postman. So perhaps our enduring, deep-rooted fear of these animals has a genetic basis. At the same time, hyenas are more like humans than many people think, being highly social animals with a fondness for meat. Small wonder, then, that humans and hyenas never learnt to coexist after the great wave of migration to Europe and Asia some 40,000 years ago. The world wasn't big enough for both of them.

Older than she looks

Although Zlatý kůň's skull aroused some interest among archaeologists in the 1950s, that soon faded. According to its initial dating, the find seemed relatively unspectacular – one of hundreds from around 12,000 years ago. This was a colossal mistake, as we now know.

The outer shape of the reconstructed skull pointed us in the right direction: to an early *Homo sapiens* who lived later than most Neandertals, but still within the age of the so-called Cro-Magnon human. The woman had – with all due respect to the dead – an extremely tough, dense skull and very thick bones to boot: all, physical features that tended to be found in very early modern humans. To this extent, traditional methods came very close to identifying what archaeogeneticists were to confirm more than seven decades later. But radiocarbon dating techniques led the researchers astray and Zlatý kůň wound up in a museum in Prague, just one exhibit among many (see Figure 8).

And yet radiocarbon dating, which had established itself in the 1950s as the standard technique for determining the age of archaeological finds, was just as revolutionary in its day as genetic dating was to become later. The way it works is by analysing carbon and its decay process in order to determine the age of bone. Dating the skull to 12,000 years ago flatly contradicted its morphology, that is, the composition of the bone, but this was never questioned at the time.

The reason for this obvious contradiction, which rendered the skull scientifically unusable for so long, is now clear: after discovery, it was reconstructed using a bone glue that was itself made from cattle bones. Thus, when it came to measuring the carbon decay, very recent animal-derived carbon got mixed up with Zlatý kůň's. This produced new and widely varying results each time – but never the roughly 47,000 years that were finally arrived at in 2021, through DNA analysis.

Up to that point, other finds were thought to constitute the oldest relics of modern humans in Eurasia. One was the thighbone of a man who died about 45,000 years ago, in Ust'-Ishim in Siberia, and another

Figure 8 What European hunter-gatherers may have looked like
© Tom Björklund

What European hunter-gatherers may have looked like – and perhaps Zlatý kůň too?

was a 40,000-year-old skull from Peştera cu Oase (the Bones Cave) near Anina in southwestern Romania. In 2020, the case of Zlatý kůň was reinforced by further evidence that the first modern humans must have spread from Africa to Europe earlier than the Oase skull had led us to believe. The bones in question, found in Bacho Kiro Cave in central Bulgaria, were dated to somewhere around 45,000 years ago, and the interesting thing about them is that they show a genetic overlap with bones of East Asians of the time. This suggests the existence during that period of a population that covered the space from East Asia to the Black Sea and interbred with the Indigenous people in both places. On the other hand, no genes from the Far East were found in the old Zlatý kůň.

At a scale of tens of thousands of years, it may seem irrelevant at first sight whether modern humans appeared on the Eurasian continent 5,000 years earlier or later; but the opposite is in fact true. This goes to the heart of one of the key questions of archaeogenetics, and indeed of our understanding of human history in general. The old data suggested that Neandertals – who probably died out some 39,000 years ago – disappeared shortly after the arrival of modern humans; hence one species displaced the other. The new datings tell a very different story, namely that modern humans coexisted and mixed with Neandertals in Europe for at least 5,000 years; only at that point did they succeed in conquering the continent.

Samples of Zlatý kůň's cranial bone were first used to study mitochondrial DNA – the genetic material of mitochondria, the part of the cell responsible for energy production. Unlike nuclear DNA, this material doesn't contain all the information found in the whole genome, but it is very useful for reconstructing lines of descent. Mitochondrial DNA is always passed on by the mother, which makes it possible to trace back the female lineage of an individual. By comparing the mitochondrial DNA of two individuals, we can calculate from the different mutations that manifest themselves – and these occur on average every 3,000 years – when the last common female ancestor lived.

In the case of Zlatý kůň, this analysis pointed to a very early split from the lineage that originally emerged in sub-Saharan Africa – the part of Africa that lies south of the Sahara – and led to the first Asians and Europeans. Only eight mutations distinguish the mitochondrial DNA of Zlatý kůň from that of the woman to whom the mitochondrial DNA of all people today who live outside sub-Saharan Africa can be traced.[1] In short, it would be hard to find anyone much closer to the original mother of the first Europeans and Asians than Zlatý kůň.

The genetic branch that led to Zlatý kůň not only diverged from the common root at a very early stage but also appears to have been very short, with no female descendants. We cannot say whether the woman had children, and we don't even know how old she was when she died. But there are no known early humans, let alone a modern human, whose DNA can be traced back directly to Zlatý kůň. Her lineage must have died out relatively quickly, so the population she belonged to couldn't have lasted long. The same goes for the humans found in Ust'-Ishim and in the Romanian Bones Cave: their genetic material is just as absent in people who live today, and their lineages have likewise disappeared. The oldest human known to date whose genes have passed down to the majority of present-day Europeans and who can thus be regarded as a direct ancestor was found at Kostenki in western Russia, where he died 39,000 years ago. He must have been a member of that successful group of modern humans whose migration proved to be a lasting achievement and coincided with the disappearance of the Neandertals – along with the gradual extinction of mammoths, other megafauna, and cave hyenas.

Like the other early modern humans found in present-day Russia, Bulgaria, and Romania and like all people outside Africa today, Zlatý kůň, too, carried between 2 and 3 per cent Neandertal DNA, as the analysis of her whole genome showed. This was no surprise, given that she came from the same population. This was, to be sure, the population that spread out of Africa and mixed with the Neandertals in the Middle East – and the population to which we owe the gene pool that exists to this day.

Atomized Neandertals

The fact that the relative proportion of Neandertal ancestry in the DNA of modern humans has remained constant over tens of thousands of years and will continue to do so can best be likened to an experiment with a glass of water and a cup of blue dye. When the population that gave rise to the first modern Europeans and Asians came into contact with the Neandertals, a drop of blue dye was added to the glass of water (metaphorically speaking, of course). When the modern humans who came out of this population went on to colonize the world, they all carried with them (to stick to the same image) a slight blue tinge, which they duly passed on to their descendants. So all subsequent interbreeding perpetuated that blue tinge by producing equally blue-tinged children. The only way to produce even bluer children would have been to re-mate with Neandertals – which did happen from time to time, for instance with the Bacho Kiro cavemen and with the ancestors of Oase Man.

While the 'blue' parts of the Neandertal ancestry in Zlatý kůň are not significantly different from our own, a few peculiarities are nevertheless observable in her genome. The main one is the length of the Neandertal DNA segments, which was markedly greater in her than in all subsequent humans. This suggests that it must have been a relatively short time ago that modern humans and Neandertals interbred, and Zlatý kůň's genetic lineage split off from the population that later gave rise to Europeans and Asians. After this interbreeding phase ended, the Neandertal content of the human genome did not decrease but became increasingly fragmented, being recombined in the DNA of each new generation.

The same process continues to this day, so that we can see Neandertal DNA dispersed throughout our genome like specks of paint on the floor of a newly decorated house. In the case of Zlatý kůň, though, we are talking whole blobs of paint. And this is how we can deduce her age: the interval separating her from the population of interbreeding Neandertals and modern humans must have been approximately seventy generations. Since the initial hybridization phase of the two human species in the Middle East ended 50,000 years ago, Zlatý kůň is estimated to be around 47,000 years old.

The desert on chromosome 1

Zlatý kůň's temporal proximity to the Neandertals is no big deal in the context of other finds. After all, we have long known of individuals who were in some cases direct descendants of Neandertals and modern humans, of Neandertals and Denisovans, or even of modern humans and Denisovans. Even Oase Man from Romania had 10 per cent Neandertal DNA, for example, and might therefore have been a great-grandchild of these prehistoric humans. All these hybrids had one thing in common with Zlatý kůň: their genetic lineages were lost. This naturally raises the question whether their fate could have been due to their genetic makeup: could the higher proportion of early human genes have proved a disadvantage in the fight for survival? Zlatý kůň can't really give us an answer to this question. But parts of her DNA certainly invite speculation. And it is through speculation that archaeology and archaeogenetics, like science in general, inch their way towards the truth.

In this spirit, another brief excursion is appropriate here. If we assume that roughly half of all Neandertal genes survive in the genomes of people living today, this means that the other half have been filtered out over the course of evolution. This may happen where a particular gene or base combination leads to evolutionary disadvantages. The possible exclusions are manifold, and there is, accordingly, an unlimited number of conceivable genetic predispositions that could dictate which species prevails and which individuals in a species come to dominate the gene pool: these predispositions can range from physical fitness to intellectual capacity.

Take the peacock, for which the absence of a showy, colourful plumage means no sex: what peacock could have dreamt, hundreds of thousands of years ago, that his muted grey shimmer would spell the end of his genetic line, while the puffed-up parvenu next to it would climb to the top of the evolutionary career ladder? It is only today that this path appears to us as the only logical one, because we cannot imagine a world of grey peacocks. In other words, the factor of chance is central to evolution. And this makes it virtually impossible for archaeogeneticists to state with any certainty precisely which genetic mutations paved the way for our ancestors' evolutionary success.

So let's go back to Zlatý kůň. We know nothing of her personal vanity, and she certainly didn't have feathers. But we can analyse the composition of her chromosome 1, the largest in the human genome. In all modern humans, this chromosome is characterized – at least from the Neandertal researcher's perspective – by a vast desert: a huge area of the genome that contains no Neandertal DNA. This suggests that, over the course of evolution, Neandertal genes were filtered out of this area on chromosome 1 and only modern human genes survived. Even our earliest ancestors – the people who emerged from the Cro-Magnon culture in Europe 40,000 years ago, with their cave paintings, Venus figurines, and musical instruments – have this desert. The desert is missing, however, in Zlatý kůň, whose only surviving legacy is a hyena-gnawed skull and possibly a few primitive stone tools. Her chromosome 1 does not have a large region without any Neandertal DNA.

Was this the reason why Zlatý kůň's genetic line petered out? And does this mean that chromosome 1 in the Neandertal was responsible for his crucial disadvantage vis-à-vis the modern humans who went on to conquer the world? Once again, we can only speculate. After all, chromosome 1 is by far the largest of the twenty-two chromosome pairs in the human genome (not counting the sex chromosome, XX in women and XY in men). With almost 250 million base pairs located on chromosome 1 – that is, roughly 8 per cent of the human DNA – it is difficult to pinpoint which genetic information could have been responsible for our ancestors' evolutionary superiority over the Neandertals. It is perfectly possible that this difference manifested itself in another chromosome altogether, perhaps in a hitherto completely overlooked position – or in the interaction between various base pairs in different places on the genome.

What we can rule out, however, is any notion that the difference between modern humans and Neandertals is due to a single gene. This assumption was still very popular in archaeology and evolutionary genetics in the noughties, but has since proved oversimplistic. The genes in the Neandertal desert on chromosome 1, for example, do not have one particular function each, but usually thousands, depending on which other genes they interact with.

In short, we should not be overly optimistic about our chances of tracking down the gene locations responsible for the Neandertals' demise by growing blobs of their brain tissue. The difference between these two human species that are so closely related is unlikely to prove observable in metabolic processes at the cellular level alone. After all, the famous 'divine spark' that made our ancestors rulers of the world is hardly something you could reconstruct in a Petri dish. It is true that our unique capacity for human culture will be rooted partly in our genes, and of course partly in the structure of our brain. But whether we will ever be able to get to the bottom of this molecular interaction remains to be seen.

No philosopher's stone

The difficulty of pinning down the phenomenon of 'culture' starts with the definition of the concept itself. There is no doubt that Shakespeare, Beethoven, the pyramids, or any creative achievement that is born of the human brain and has the capacity to stimulate pleasure in others can be described unequivocally as culture; but then the same could be said of soap operas, cock fights, and reality shows. Different people are likely to associate completely different things with the term, which goes to show just how elusive it is. Think of a vegan and a steak lover discussing food culture, a pop fan and a classical music buff exchanging playlists, or a football fanatic raving on to an indifferent listener about the world-transforming importance of his substitute religion.

We should, moreover, beware of exaggerating the importance of human culture, given that it also encompasses – and always has – the worst crimes of humanity; in fact the two things are often inextricably intertwined. The pyramids, which inspire such awe in us today, were built on corpses of untold thousands of workers; and some artists whose work we now praise to the skies subscribed to barbaric ideologies that led to the death of millions of people. Human culture, then, is essentially a neutral term, denoting the source of everything that humans have brought to the planet, for better or worse, over the course of their history.

The artistically carved flutes, the Venus figurines, the cave paintings: all these are unquestionably signs of a complex culture that came to Europe some 40,000 years ago, when the flame it probably sparked somewhere in Africa more than 80,000 years ago spread throughout the globe. This rapidly developing artistic skill was just one feature of a fundamental behavioural change that humans, or rather humankind, must have undergone. During this period we seem to have discovered that there is more to life than the next warm cave and the next meal.

Survival began to give way to living. Once the first human had the idea of honing a bird bone so that it produced a sound and, with a bit of practice, a tune, this may have had a domino effect, producing increasingly sophisticated achievements. Ultimately there were probably quite mundane factors at play here too, as in the case of the peacock and other animals: a talented flute player, just like a good hunter, may have greatly improved his chances of procreation through his skill. Thus driven, modern humans made rapid strides out of the Stone Age and towards an advanced civilization.

The shell seekers

Not so the Neandertals. For decades, dedicated researchers around the world have been avidly searching for remains of this extinct human type that might point to a complex Neandertal culture; but to little avail. While these early humans were also capable of producing tools and making fire, there is no evidence to date of any intricate skills or any Neandertal art.

That said, in 2010 archaeologists came across some 50,000-year-old shells with perforations that suggest that they might have been worn by Neandertals as jewellery. The finds, which were made in two caves in south-eastern Spain, contained traces of coloured minerals that were thought at the time to be possible relics of Neandertal cosmetics. But – and this is a possibility – the holes could simply have been the result of 50,000 years of erosion, and the pigment could have found its way there by other means. In which case the finds would not be art but a caprice of nature.

On the other hand, it cannot be an accident that the 40,000-year-old artefacts found in the Lone Valley caves of Swabia evoke a lion man (Figure 9) or a full-figured woman, nor could it have been a natural process that made six fine, equally spaced holes in a bird bone. The fact that we have so many artefacts from the Cro-Magnon era also has a lot to do with the burial rituals established at that time. There are many graves in which the dead were not simply buried but laid to rest amid grave goods, notably jewellery and weapons, as well as other everyday objects.

Figure 9 'Lion Man' © Tom Björklund

Stone Age artefacts such as 'Lion Man' from the Lone Valley in Swabia, Germany, still command admiration even today. Could this one also have served as a children's toy?

There are no signs of such complex burial rituals in the case of Neandertals, despite the occasional claim – based on chance discoveries of stone flakes next to Neandertal bones – that they, too, buried gifts with their dead. In reality these are just as likely to have been stone tools mislaid in the vicinity. No Neandertal graves have been found in which the dead were deliberately placed in a particular way – recumbent, seated, or facing another body – whereas the Cro-Magnons have given us dozens.

Most of the Neandertal bones studied by modern archaeogeneticists and archaeologists are from sites where someone died on the spot and was then presumably devoured by hyenas and other scavengers – a sight that our ancestors wished to spare themselves by burying their nearest and dearest. Perhaps burial also helped them to suppress the painful awareness of their own membership of an animal kingdom that they had yet to fully subjugate. Giving their relatives a dignified burial may have served to reassure them of their emancipation from the cruelties of the eternal cycle, not just in life but also in death. Finally, burying their own was a way of ensuring that they would be afforded the same protection by their group after their own death, rather than ending up as some passing creature's supper.

Homo sapiens *acquires self-consciousness*

Nevertheless, at this stage we are still worlds away from the 'community of values' that developed with modern humans, encompassing all individuals and fostering a unique sense of fellowship and solidarity. No matter how impressive their cultural zeal, their reverence for life, and the social cohesion within their own groups, humans could equally be cruel towards other groups. In this respect, there was no difference between modern humans and Neandertals in either the nature or the intensity of their violence. And they probably resorted to it almost always when two groups encountered each other on the sparsely populated Eurasian continent and had to decide on the spot whether to pre-empt a possible attack by the enemy by going on the offensive.

Be that as it may, with the first cultural skills, modern humans developed a sense of themselves as enjoying some special status within the cosmic order. This awareness has borne humanity along to this day, and there is much to suggest that genetic differences must be partly responsible for its superiority. This divergence occurred in the modern human lineage after it split, along with the Neandertals, from our common ancestor in Africa and our ninety fixed differences evolved. Could it be, then, that one of these mutations even produced a mild form of madness whose genetic foundations spread at lightning speed, giving rise to a population that could not but create a deeper meaning for the aimless wandering that constituted human life at that time – through mysticism, music, and cave painting?

Should a genetic explanation be found one day for the victory march of modern humans – a march that led to their conquest of the world 40,000 years ago, leaving the Neandertals and Denisovans far behind – it will undoubtedly lie in Africa. Here we must challenge the traditional anthropological narrative that still prevails to this day, namely that *Homo sapiens'* dispersal from Africa via Eurasia to the rest of the world was a complete walkover. In reality it was achieved only after many false starts and setbacks. The north, where Neandertals and Denisovans had found their niche, was unattainable for our ancestors. The qualities responsible for their ultimate success evolved in the Cradle of Humankind: sub-Saharan Africa. This was the home of the genetic melting pot that gave rise to *Homo sapiens*. And *Homo sapiens* was just one of many candidates for the crown of creation.

3

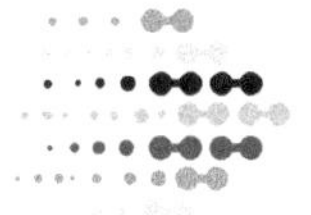

Planet of the Apes

Time for a flashback to millions of years ago, when it all began. Africa is teeming with all kinds of great apes, the majority of which end up in an evolutionary cul-de-sac, or buried in the Alps, like Udo. On their way out of this melting pot, humans begin to push into Europe. The end of the line is Greece – for now.

Territorial claims in the Middle East

The place where modern humans who were spreading out from Africa came up against the dominance of Neandertals remains to this day a flashpoint of irreconcilable territorial interests, albeit under completely different circumstances. It must have been in the Middle East, and specifically in the region of modern-day Israel, that the habitats of these two early human species overlapped: a scene of constantly shifting boundaries and a point of departure for occasional expeditions to the north by modern humans whose lines subsequently died out, as in the case of Zlatý kůň. This was a prelude to the great trek northwards that the ancestors of all non-Africans were later to embark upon, most likely from the Arabian Peninsula.

That migration unfolded over millions of years, during which humans learned to walk upright, developed a high-powered brain, and cultivated cultural techniques. The long run-up to it took place in Africa, but not in one particular location, as most researchers assumed until a few years ago: it was probably dispersed around that vast continent. It was this melting pot of diverse human lineages that eventually gave rise to the Africans, who became the ancestors of all people alive today. Until that

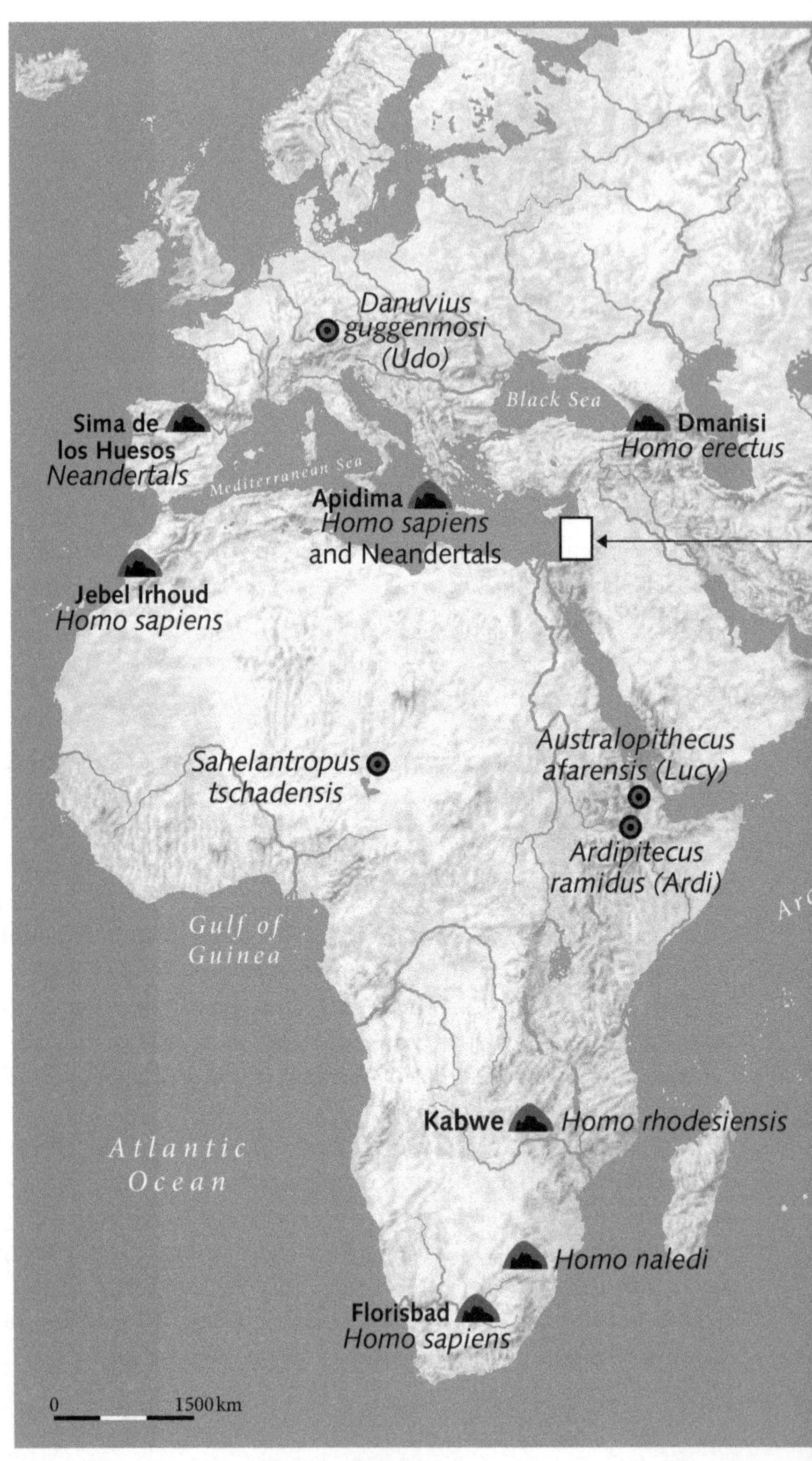

Figure 10 Planet Ape.

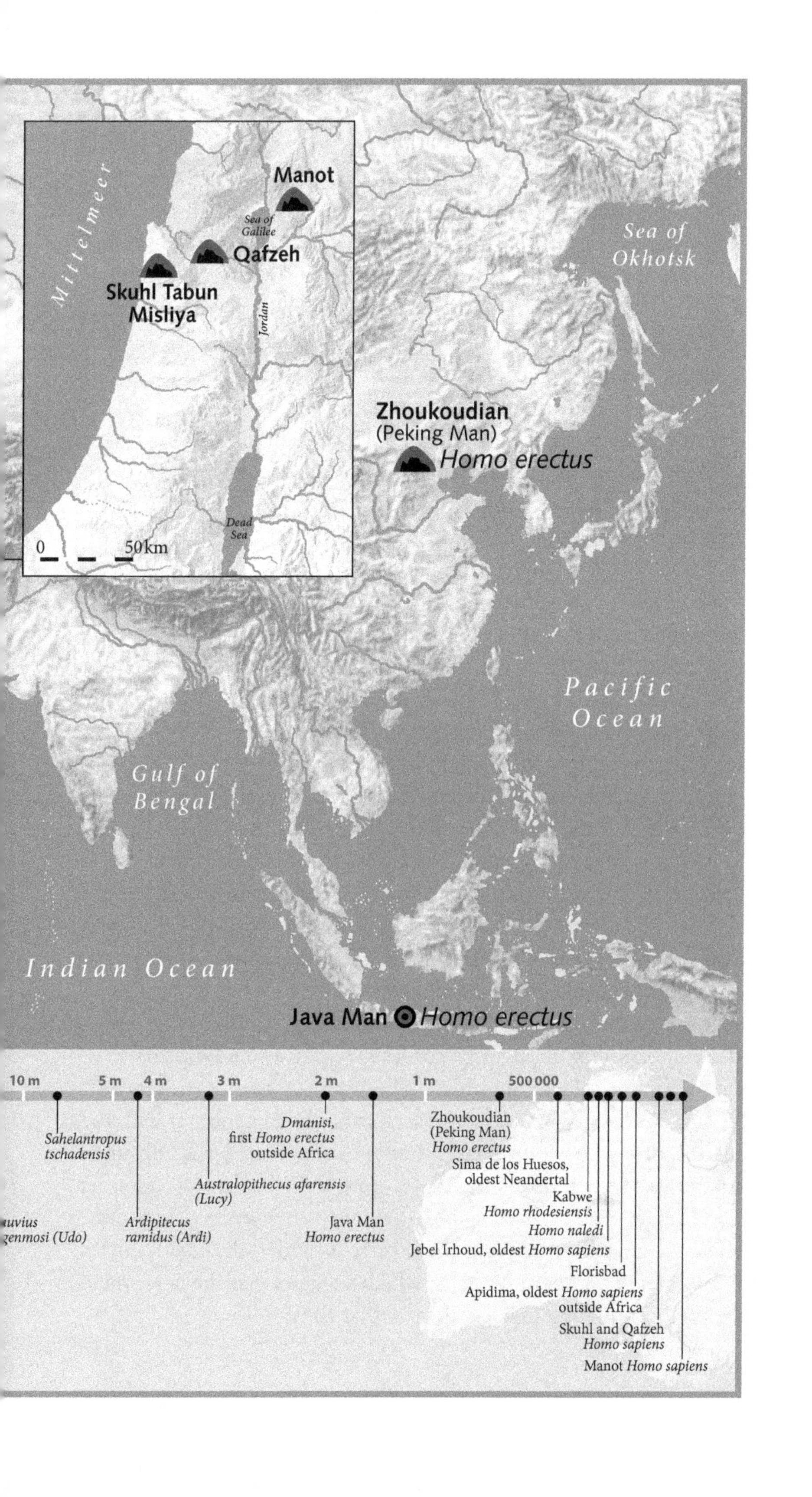

Manot
Sea of Galilee
Qafzeh
Mittelmeer
Skuhl Tabun
Misliya
Jordan
Dead Sea
0
50 km
Sea of Okhotsk
Zhoukoudian
(Peking Man)
Homo erectus
Pacific Ocean
Gulf of Bengal
Indian Ocean
Java Man Homo erectus
10 m
5 m
4 m
3 m
2 m
1 m
500 000
Sahelantropus tschadensis
Ardipitecus ramidus (Ardi)
Australopithecus afarensis (Lucy)
Dmanisi, first Homo erectus outside Africa
Java Man Homo erectus
Zhoukoudian (Peking Man) Homo erectus
Sima de los Huesos, oldest Neandertal
Kabwe Homo rhodesiensis
Homo naledi
Jebel Irhoud, oldest Homo sapiens
Florisbad
Apidima, oldest Homo sapiens outside Africa
Skuhl and Qafzeh Homo sapiens
Manot Homo sapiens

point, humans left nothing untried in their quest to conquer the north. Signs of such attempts can be found in their hundreds, from Greece to China: a series of shattered dreams in the gallery of human history.

From northern Israel, from the eastern shore of the Sea of Galilee and up to the coast below Haifa, we now have multiple fossils, both of resident Neandertals and of early modern humans. Some of these sites are less than a stone's throw from each other. Even if the individuals found there were not contemporaneous, the accumulation of finds in this part of the world suggests that the mingling of the two species manifested in our present-day DNA took place there. It was a coexistence that will certainly have entailed bloody battles for food, resources, and sexual partners.

Numerous skull finds attest to the existence of Neandertals in the region of present-day Israel; they were indigenous to it until their extinction. There are fewer fossils of modern humans from that era, but they do exist. Meanwhile new bones have been found that suggest an encounter between Neandertals and modern humans as far back as 55,000 years ago. In 2015 a skull found in Manot Cave on the eastern shore of the Sea of Galilee was dated to this age. Although it was no longer possible to extract any DNA from it, the reconstructed morphology of the skull, of which only a fragment of the cranium remained, points to a modern human.

Also found in the vicinity of the bone was a prehistoric fire pit. The age of the Manot skull coincides almost exactly with the point in time when, according to our DNA, the ancestors of all present-day non-Africans appear to have picked up their Neandertal DNA. Hence this find could be one of the most important made so far when it comes to tracing the chronology of human expansion into Eurasia: perhaps the man or woman who lived in Manot Cave 55,000 years ago was among the direct ancestors of Eurasians. This cannot be regarded as anything more than a possibility, though, given the lack of useable DNA data.

Manot is by no means the oldest clue we have to the existence of modern humans outside Africa, however – or to their encounter with the Neandertals. After the evolutionary lineage of the Neandertals diverged from that of modern humans in Africa more than half a million years ago, the latter appear to have found their way to Europe. There was, however, some gene flow between those early Neandertals outside Africa and the ancestors of modern humans, which suggests that the latter must have already been living outside Africa at that time.

Our contribution to the Neandertals

After the decoding of the Neandertal genome proved the existence of a Neandertal element in our DNA, the theory of a close connection between the two hominins was further reinforced in 2016: only, this time, traces of *Homo sapiens* DNA were found in Neandertals. It was a spectacular discovery, which ultimately led to the revision of our evolutionary timetable. This achievement was possible thanks to a group of hominins who lived in Spain 420,000 years ago and whose DNA researchers managed, against all expectations, to sequence. Once again, the bones were from the highly productive site of Sima de los Huesos. Until then the Neandertal lineage was assumed not to have emerged any earlier than 420,000 years ago in Africa, which meant that it couldn't possibly have spread to the furthest corners of Europe so soon.

The DNA from Spain not only disproved this theory but revealed a great deal more. It differed significantly from the previously sequenced DNA from other female Neandertals, all of whom were also much younger. A comparison with the genomes of modern humans showed that in all sequenced Neandertals from the period after the early Spaniards an element of *Homo sapiens* DNA was present.

This effect was first observed in the mitochondrial DNA: so at least this fragment of DNA, which is always inherited from the mother, originated from a modern female human. In 2020, this analysis was repeated on the Y chromosome of male Neandertals – that is, on the sex chromosome passed on from father to son. Here, too, the same result was obtained as for the mitochondrial DNA: the Y-chromosomes derived from an early modern human lineage. In other words, at least two early modern humans – a male one and a female one – contributed to the gene pool of most Neandertals, and probably to far more than that.[1]

Thanks to the gene clock, split-read analyses, and DNA comparisons, we can thus infer that a gene transfer must have already taken place between early modern humans and Neandertals 420,000 years ago at the earliest and 220,000 years ago at the latest, as subsequently reflected in the Neandertal gene pool.

The (very) ancient Greeks

In 2019, a team from Tübingen University analysed two skulls that had been discovered more than forty years before in Apidima Cave on the Mani Peninsula of southern Greece but that researchers had found little use for at the time. The idea of re-examining the skulls using new methods was, as it turned out, a good one: as a result, we now have not only a new oldest known modern European, but also a new hotspot where Neandertals and modern humans may have come into contact with each other.

The fragments from the two skulls are known as Apidima 1 and Apidima 2. Both are poorly preserved and very hard to reconstruct – particularly Apidima 1, which, being half-buried in a rock, would probably not even be recognizable as a skull bone to the untrained eye. Using computed tomography and other methods, the researchers reconstructed the shape of the skull from its fragmented components. Its age was determined via a technique known as uranium–thorium dating, which measures processes such as stalagmite decay in order to date cave finds contained in layers of rock.

In the case of the rock-encased Apidima 1, a minimum age of 210,000 years was established, while Apidima 2 was dated to 170,000 years ago. At the same time, the skull shape of Apidima 1 exhibited features of a modern human: coupled with the dating, this was sensational. Up until then, the first modern humans were not thought to have reached Europe until the great migration more than 45,000 years ago. And even though Apidima 1 shows both archaic and modern features, after complete reconstruction and comparison with other early Europeans, there was no longer any doubt that it belonged to the modern human species. Of its genes we know nothing, as the samples did not contain any sequenceable hereditary material. We can, however, rule out Apidima 1 as a possible ancestor to modern Eurasians: it is far too old.

What is quite possible, on the other hand, is that Greece was the place where modern human DNA was transferred to the Neandertals who went on to dominate Europe. Whether – taking this to have been the case – it was perhaps even Apidima 1 him- or herself who mated with a Neandertal, we will never know, just as we will never know the extent of the advance evidently accomplished by these early Greeks. Was it merely a foray by a bold but insignificant tribe, something that lasted no more than a few generations, or was there in fact a hunter-gatherer culture in the southern Peloponnese that became stabilized over the millennia?

The latter possibility is probably the least likely. While we can now point to hundreds of Neandertal finds, the same cannot be said for modern humans outside Africa. Any bone dating back beyond the magic threshold of 45,000 years is still headline material, at least in academic journals. For one of the many Neandertal bones, on the other hand, we need to look no further than Apidima Cave itself, as we know from the 2019 analysis that Apidima 2 belonged to a Neandertal. This individual inhabited the cave 40,000 years later than Apidima 1. If the Neandertals had settled here before that date, however – and there is little to contradict such a theory, given this prehistoric type's long history of colonization in Europe – then this region might have been the place where the two species first interbred.

Greece is just one possible site of this event; of course, it could equally well have taken place in what is now Israel. Back in the early twentieth century, archaeologists discovered the bones of several individuals in Skhul Cave in Mount Carmel, northern Israel: these were dispersed across various burial sites and belonged to modern humans who are estimated to have lived there about 120,000 years ago. Bones from Qafzeh Cave, some 40 kilometres east of Skhul, were also dated to roughly the same period. When archaeologists unearthed these fossils in the 1930s, they interpreted them as representing a transitional species between the Neandertal and the modern human, in accordance with archaeological orthodoxy at the time.[2] Today we know that Skuhl and Qafzeh were in fact modern humans, but with an archaic skull shape. The humans from Skuhl and Qafzeh are not among our ancestors either, having died out long before the Neandertals, probably at least 80,000 years ago.

For decades, Skuhl and Qafzeh were regarded as the oldest remains of modern humans outside Africa. Before 2019, when Apidima 1 shifted this milestone much further back in time and notably further northwards, there was another groundbreaking discovery that pointed to the Middle East's central role in human history. Like the Skuhl site, Misliya Cave lies in the Carmel range, just ten kilometres further north. On the basis of evidence from an upper jaw bone complete with eight teeth, this cave appears to have sheltered modern humans as long as 180,000 years ago. And it comes as no surprise that the Neandertals, too, favoured the Carmel range: remains from them have been found just a few metres from Skuhl, in Tabun Cave for example, where they lived about 120,000 to 80,000 years ago.

Short, stocky, superior

There is no doubt, then, that modern humans and Neandertals in the region of present-day Israel used the same dwellings – in fact the market for caves seems to have been almost crowded at times. It is unlikely that the two species lived together, though: the boundaries of Neandertal and modern human habitats were probably constantly shifting as a result of often extreme climatic changes. Whenever it grew warmer, modern humans may have inched further north, from the south of the Arabian Peninsula into present-day Israel, for example. Our ancestors had, after all, evolved to cope with a hot climate and the Neandertals with the colder north – partly thanks to their squat and stocky body type, which was better able to store heat.[3] For their part, the Neandertals would undoubtedly have had no problem with the higher temperatures, but they were entirely dependent on the steppe and the megafauna it sustained; with rising temperatures, these hunting grounds would have shifted northwards.

The Pleistocene, which began 2.6 million years ago, brought countless temperature variations that offered ample opportunity for this north–south movement of ecosystems. Warm phases such as the Holocene – which began 12,000 years ago and is still ongoing – occurred periodically in the past, albeit without the supercharging effect of modern human activity, of course. In extreme cases, these phases saw global temperatures rise to up to 2 degrees Celsius above the present average. And even during the cold phases there were major upward variations, when temperatures could be similar to those seen in the cooler phases of the current warm period.

The majority of early modern human finds north of Africa date from eras when there were perceptible climatic upticks. The longest and hottest 'interglacial' was the Eemian, which, give or take, began 126,000 years ago and ended 115,000 years ago. This period is named after the Dutch river Eem: there, in the nineteenth century, the Rotterdam-based doctor and geologist Pieter Harting discovered snail and shellfish deposits that are now found mostly in the Mediterranean. At the peak of this warm period the average global temperature was higher than today, and rhinos and hippos roamed through what is now the city of London. This is when the modern humans lived whose remains were found in the Skuhl and Qafzeh caves.

By contrast, the lifetime of the oldest modern human outside Africa known to date, Apidima 1, falls in the middle of the Saale glaciation (so named because the Ice Age glaciers reached to the river Saale in Thuringia, Germany). During the peaks of this cold period, average global temperatures could fall as low as 9 degrees Celsius, making conditions on the Eurasian continent inhospitable to humans who had come from Africa. And yet the fact that at least Apidima 1 lived in southern Greece during that period is not contradictory, because in his lifetime the Saale glaciation was interrupted by what is known as the Wacken interglacial, when average temperatures rose significantly. During that spell, conditions would have been mild in what is now Greece.

On the other hand, Neandertal Apidima 2 would have experienced a very different climate during his lifetime – that is, around 170,000 years ago. With the exception of the Last Glacial Maximum, this was one of the coldest periods in the past 2 million years. The long era in which Neandertals lived in Mount Carmel's caves is also consistent with the logic of temperature-driven habitat shifts: 115,000 years ago, when the warm Eemian period ended, this region clearly became much too cold again for modern humans. It must have taken them ages to settle here again – a theory supported by the 55,000-year-old skull from Manot Cave.

Whether Neandertals and modern humans came into conflict during the eventful history of prehistoric Israel remains pure speculation in the absence of any finds to testify to their coexistence. But it is not unlikely. While most of the Arabian Peninsula boasted lush vegetation during the Ice Ages, it certainly won't have offered unlimited hunting and thus could well have been the scene of violent clashes – most likely over prey, but no doubt also over the choice of a mate. The Neandertals, who had an unassailable advantage as experienced hunters of mammoths and other Eurasian megafauna and were also more familiar with the dangers that lurked there, probably perceived their new rivals more as a nuisance than a mortal threat, as the scattered traces of early African migrants stand proof.

It may be that all modern humans who attempted persistently, over tens of thousands of years, to expand their habitat were bold adventurers. Or perhaps it was sheer necessity that drove them northwards. All we know is that their genetic footprints were lost. These perpetual false starts had become part and parcel of the culture that our ancestors developed in Africa and that eventually laid the foundation for a successful expansion. But that journey was a long one; it started with the very first apes that came down from the trees – eyed suspiciously, perhaps, by others of their kind, for which this may have been a step too far.

Udo the upright ape

Before our ancestors could set out to conquer the planet, they had first of all to emancipate themselves from our closest relatives, the great apes. As we know from genetic calculations, the evolutionary line that leads to modern humans, Neandertals, and Denisovans diverged around 7 million years ago from the ancestor we share with the chimpanzee. The forebears of chimpanzees, humans, and gorillas lived about 10 million years ago, the orangutan line having split off some 15 million years ago. In short, great apes were present throughout Africa and Eurasia for millions of years, long before humans appeared on the scene.

The Miocene epoch, which began 23 million years ago and ended 5.3 million years ago, offered ideal conditions for the expansion of great apes. Both Africa and Eurasia boasted a tropical climate, and the north was covered with rainforest and jungle. How many species existed back then can no longer be determined: in such conditions, a dead great ape would have been eaten and broken down by bacteria in no time, leaving almost no fossils behind. But in all probability there were at least a dozen different species of great apes that spread from Africa to Europe and East Asia. Among these was Udo, who lived in the Allgäu region of southern Germany 12 million years ago (Figure 11).

Officially known as *Danuvius guggenmosi*, this great ape, found in a Bavarian claypit, owes its nickname to the fact that parts of its skeleton, including an almost complete shinbone, were discovered on 17 May 2016, which happened to be the seventieth birthday of German rock

Figure 11 Udo the great ape © Velizar Simeonovski

An artist's reconstruction of Udo, the great ape who lived in the foothills of the Alps 12 million years ago. But whether he was really able to walk upright, as was assumed after his discovery in 2016, remains a matter of debate.

legend Udo Lindenberg. While reconstructing the bone fragments, the Tübingen-based (and evidently rock-loving) team of researchers made a surprising discovery: Udo could, it seemed, walk upright. He thereby possessed an ability that great apes were previously thought to have acquired only 7.5 million years ago, and in the African steppe rather than the foothills of the Alps. This sparked feverish speculation – among journalists more than among scientists – as to whether Udo could have belonged to a precursor population of upright humans and whether the cradle of humanity therefore lay in Central Europe rather than Africa.

This is highly unlikely. For one thing, such a hypothesis implies that Udo's descendants would have given up their upright posture on the way south. Furthermore, a highly fragmented skeleton dating back millions of years is a somewhat unreliable basis for judging whether this individual really did walk upright and, if so, whether he did it only occasionally or as his normal gait. What Udo does prove, however, is just how widely our relatives, the great apes, had already spread around the world all those millions of years ago. They, not us, were the first to leave Africa. And for a long time they were far more numerous than the later human arrivals.

This we can tell from the DNA of the great ape species still in existence today. With the aid of genetic analyses, it is possible to calculate the size of the populations from which all present-day individuals are descended. Chimpanzees, for example – who are native to Central Africa and currently number around 200,000 individuals – can be traced back to an original population of roughly 50,000, meaning that all chimpanzees derive their gene pool from this group. In the case of gorillas – of whom there are some 360,000 – the equivalent figure is roughly 40,000 original individuals. By comparison, the genetic diversity of the 8 billion humans is relatively small, being traceable to a population of no more than 10,000 individuals who lived about 300–400,000 years ago.

It is perhaps just as well that those humans didn't leave their native home any earlier, as they might otherwise have run into some rather scary members of the great ape family: the *Gigantopithecus* for one – an offshoot of the orangutan that lived up to 300,000 years ago in East Asia and reached a height of up to eleven and a half feet. That said, it is possible that at least *Homo erectus* encountered *Gigantopithecus*, having inhabited the same region a million years ago in the form of Peking Man and Java Man. Even as recently as 10,000 years ago, great apes probably still outnumbered modern humans.

When the Mediterranean disappeared

Five million years ago there was a climatic change that, while it didn't yet overturn the balance of power, certainly created ideal conditions for the emergence of upright early humans in Africa. At that time the Miocene gave way to the Pliocene, rendering the landscape far less congenial to apes, both in Eurasia and Africa: from now on the climate began heading towards the Ice Age that was to begin three million years later.

During the Pliocene, parts of the Arctic began to freeze over, as the Antarctic had already done in the late Miocene. As a result, more water was retained at the poles and the global climate became increasingly dry. This even caused the Mediterranean Sea to dry up during the transition from Miocene to Pliocene. Known as 'the Messinian salinity crisis', this event saw the Mediterranean lose its connection with the Atlantic as a result of the drop in sea level triggered by the glaciation of the polar regions and the effects of continental drift. Evaporation increased and, as in the Dead Sea, the water level continued to fall. It was not until the land bridge between Europe and Africa began to sink slightly, 5 million years ago, that water from the Atlantic could flow back into the vast basin. Eventually the growing flow pressure gave rise to the Strait of Gibraltar. Through this channel, the Mediterranean then filled up rapidly, forming since then a natural barrier between North Africa and Europe that remained insurmountable for humans for a very long time.

It wasn't just in Eurasia that the ape-friendly jungle receded; it did so in Africa, too. First it was displaced in the north of the continent by savannah grasslands that expanded increasingly with the cooling climate, coming to cover parts of sub-Saharan Africa by the end of the Pliocene. From then on, the upright gait became a passport to a new world for all species of great ape that mastered it. This evolutionary revolution was probably what turned humans into hunters. Indeed, no other mammal can run continuously over such long distances as the human. This meant that humans could spot their prey from afar, thanks to the elevated position of their head, and then simply pursue it to the point of exhaustion before killing and disembowelling it with primitive weapons.

It was this new source of protein that allowed the development of a brain that consumes around a quarter of the human body's energy while accounting on average for just 2 per cent of its weight. Thanks to this organ, humans were able to survive in the steppe, by developing either new hunting strategies or better weapons and techniques for obtaining plant-based nutrients from the grasslands – for example for cracking nuts and shells open, or for digging up bulbs and roots.

The upright gait evolved not in the steppe but in the jungle, most likely as a tentative experiment – literally a step-by-step process. Some species developed the ability to walk upright, but generally moved around on all fours, ready to shin up a tree at the first sign of danger. Even after the lineages leading to us and to the present-day chimpanzee diverged, it was a long time before the two species were able to keep their hands off each other: for around 1.5 million years, genetic exchanges continued to take place periodically between these great apes until they were no longer biologically possible, as we know from present-day chimpanzee DNA.

The first, though by no means unequivocal signs of the presence of upright great apes in Africa are about 7 million years old. At that time, the shores of present-day Lake Chad were home to the species *Sahelanthropus tchadensis* – the name given to what some researchers consider to be the oldest fossil in the human lineage. But given its age and poor condition, there is some room for interpretation, to say the least. By contrast, the genus *Ardipithecus*, which lived 4.4 million years ago in present-day Ethiopia, can be said with some certainty to have been a human forerunner, despite bearing a much closer physical resemblance to an ape than to us: its posture was more bent than upright, though it did walk on two legs, while the shape of its hands and feet suggests that it remained an enthusiastic climber. For this reason, it cannot yet be described as human. The same goes for what was probably our most famous ancestor from the era of transition from ape to human: Lucy. She lived 3.1 million years ago, in Ethiopia, and belonged to the genus *Australopithecus* – just like the Dikika baby, an approximately three-year-old *Australopithecus afarensis* excavated also in Ethiopia (Figure 12).

The australopithecines survived for an exceptionally long time – over two million years. Although they were precursors to the first humans, their particular branch led eventually to an evolutionary dead end. Then came the species *Paranthropus robustus*, which inhabited sub-Saharan Africa 1.8

million years ago. It lived up to its name, having a squat body and huge teeth, which were evidently needed to crush nuts and plants, as *Paranthropus robustus*, unlike Lucy's relatives, was a herbivore rather than an omnivore. This proved to be its evolutionary doom: this species was sidelined and disappeared forever from the scene around 1.5 million years ago.

This was not the fate of the genus *Homo* – another descendant of the *Australopithecus* – which emerged around 2.5 million years ago and to which we of course belong. It has not been definitively established whether *Homo rudolfensis*, which lived 2.5 to 1.9 million years ago, or *Homo habilis*, thought to have existed 2.1 to 1.5 million years ago, were the actual forebears of *Homo erectus*, but this hardly matters in the bigger

Figure 12 *Ardipithecus*, *Australopithecus*, *Homo habilis*, and *Homo erectus*
© Tom Björklund

Africa is the cradle of humanity. Before the arrival of *Homo sapiens*, countless early humans traversed the continent, but only a few have left traces behind: those we know of include *Ardipithecus*, *Australopithecus*, *Homo habilis*, and *Homo erectus* (from top right to bottom left).

picture. The point is, both *Homo* species refined the technique – previously learnt by Lucy's kind – of using split stones as tools. It was then, too, that the human brain began to grow, being already larger than that of *Australopithecus* by a third – at 600–750 millilitres.

Since we have no DNA from *Homo habilis* or *Homo rudolfensis* but only bones, we cannot say whether there was any genetic mixing between the two populations. However, all the archaeogenetic insights of recent years into the intermingling of all possible early humans and great ape families suggest that both *Homo rudolfensis* and *Homo habilis* might have been able to mix, as they even inhabited overlapping areas of East Africa and were quite closely related.

The ultimate melting pot

Roughly two million years ago, *Homo erectus* finally appeared on the scene. This was the evolutionary leap that first brought humans to Eurasia. Unlike their immediate predecessors, who walked upright only intermittently, members of the species *Homo erectus* travelled far and wide on two legs. It is a matter of debate whether they were already successful hunters, but there is no doubt that meat, or at least carrion, constituted a large part of their diet. And, given their evolutionary adaptation to long-distance running (for which bipedalism offers the most energy-efficient solution), there is good reason to believe that they were masters of endurance hunting.

Homo erectus was and remains to this day the most successful human species, surviving far longer on the planet (around 1.5 million years) than modern humans can yet claim. During that time, the brains of its countless subforms grew to a capacity of up to 1,000 millilitres. This compares with an average capacity of around 1,200 millilitres in modern-day people, or as much as 1,500 among the Neandertals.

The upright human clearly lost no time in applying his/her new intelligence and stamina: according to the oldest surviving relic of human life outside Africa, which was found on the Dmanisi Plateau in 1991, this species lived in the Caucasus as long as 1.8 million years ago. To date, hundreds of *Homo erectus* fossils have been unearthed around the globe. The Ice Age did not, it seems, prevent these early humans from migrating northwards. Even so, their star faded too, albeit after a lengthy period.

Initially Eurasia was the sole preserve of Neandertals and Denisovans, who, like modern humans, split off from a lineage of *Homo erectus* in Africa.

At this stage there was no hint of the modern human yet – the 'rational', 'thinking' *Homo sapiens* – in Europe or Asia. But where exactly in Africa did the species first appear? Which genetic lineage led to the human culture that came to stamp its identity on the entire planet? This is a question that has occupied scientists for decades, but continues to lead nowhere. The idea that modern humans emerged from an original population in some African valley is as wide of the mark as the notion that a certain mutation gave modern humans a conquering gene, or that human civilization experienced a Big Bang that catapulted the *Homo* genus from *erectus* to *sapiens*. The more likely explanation is that Africa was the first and most important melting pot, where a variety of genetic lineages combined to form our ancestors. This theory of a pan-African evolution is gaining increasing currency among anthropologists. Over the past few years, more and more candidates have been identified whose DNA may have gone into the pot. These include the ancestors of the Greek specimen Apidima 1 and those of the humans who inhabited the Misliya Cave in Israel 180,000 years ago, all of whom were alive during the era of this pan-African fusion.

That said, the oldest known remains of a modern human to date did indeed originate from Africa. This fossil – a skull – was found in Jebel Irhoud, Morocco. In 2019, a team from the MPI EVA dated it to 300,000 years ago and, on the basis of a computer reconstruction, classified it as a very early *Homo sapiens*, contrary to the accepted view that the species appeared 200,000 years ago at the earliest.

Other human forms also shared the continent with early modern humans. One of them, found in a South African cave, was labelled *Homo naledi* and had lived there up to 250,000 years ago. This find was a bizarre one for everyone concerned. The archaic appearance of the *Homo naledi* fossils – of which numerous specimens have since been found – led most anthropologists to assume at first that they were prehistoric human bones up to two million years old. But the location gave them pause: the bones lay in a cave whose entrance had apparently been sealed all along with a large stone, so that it could be accessed only with a nifty bit of climbing. This by itself wouldn't rule out an early prehistoric human. What might, however, was the existence of a metre-high bone bed, which meant that the individuals associated with it must have been dragged over the stone; so all the evidence pointed to a burial site. Then in May 2017 came the bombshell, when the bones were analysed: this vertically challenged archaic species, which was evidently capable of burying its dead, was not in fact millions of years old, but 335,000 years at the most. This was, then, another human type that could have passed on its genes to our ancestors.

The same goes for *Homo rhodesiensis.*[4] The skull of 'Kabwe 1', which is thought to belong to this species, was discovered almost a hundred years ago, in present-day Zambia, and initially estimated to be up to 2.5 million years old. Thanks to a new dating in 2020, however, we now know that it, too, is around 300,000 years old – the same age as the population whose existence was inferred from a skull found in Florisbad, South Africa, in 1936. The Florisbad find could be another fossil of an early modern human, albeit with some additional similarities to *Homo erectus* and *Homo rhodesiensis.*

All these clues point to the same conclusion: that the hundreds of thousands of years during which the DNA pool of our African ancestors is thought to have evolved were characterized by a huge genetic diversity across the African continent. At the same time, the African landscape must have been constantly changing along with the climate, so that, sporadically, the natural barriers that prevented the mixing of populations diminished, increased, shifted, and sometimes even disappeared altogether. One example is the savannah, which displaced the rainforest as the climate became dryer, thus creating new corridors for humans – quite apart from the Great Rift Valley that, running as it does right through the eastern part of the continent as far as present-day Israel, already provided a corridor that was passable, at least in part. Hence, when it comes to determining which species of early humans interbred in the lead-up to the birth of our ancestors, we can only make guesses on the strength of finds that cannot possibly cover the whole spectrum.

Before that point in our history, however, tens of thousands more years were to elapse. Only then, finally endowed with *Homo sapiens*' upright gait and understanding of the world, did the modern human begin to develop a flourishing hunter-gatherer culture and look beyond the immediate horizon, into a promised land whose inhabitants had no intention of giving it up.

4

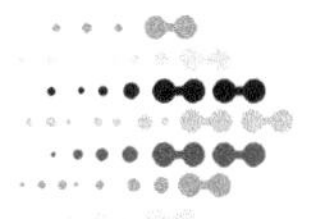

Apocalypse

About 74,000 years ago, a volcanic eruption puts a temporary end to human expansion. There is virtually no way out of Africa. Our ancestors seize the few opportunities available to them. Only once do they succeed.

Off to sunny Spain!

The wanderlust of early humans – whether *Homo erectus* or *Homo sapiens* – was boundless. Whenever possible, some group or another would try to reach the north. Not that this was strictly necessary: the vast African continent must, after all, have offered ample space and attractive hunting grounds for populations to expand. Although Africa is scarcely more than half the size of Eurasia, for humans accustomed to the southern climate and its flora and fauna, only small parts of the northern hemisphere were habitable during the Ice Age: the Middle East, the Indian subcontinent, and Southeast Asia. The oldest known non-African *Homo erectus*, who lived in the Caucasus 1.8 million years ago, had therefore ventured very far north for that time, perhaps taking advantage of a temporary mildening of the Ice Age. The region most frequented by *Homo erectus* over the millennia, however, was the southern corridor along the Indian Ocean, where regular discoveries of new bones point to a very early expansion of our forerunner model into Southeast Asia.

These early human populations extended from western Eurasia far into the east: *Homo erectus* inhabited the Iberian Peninsula as early as 1.2 million years ago and would certainly not have arrived there via the Strait of Gibraltar, but rather via the Middle East, probably along the northern shores of the Mediterranean. This species had nearly reached the furthest

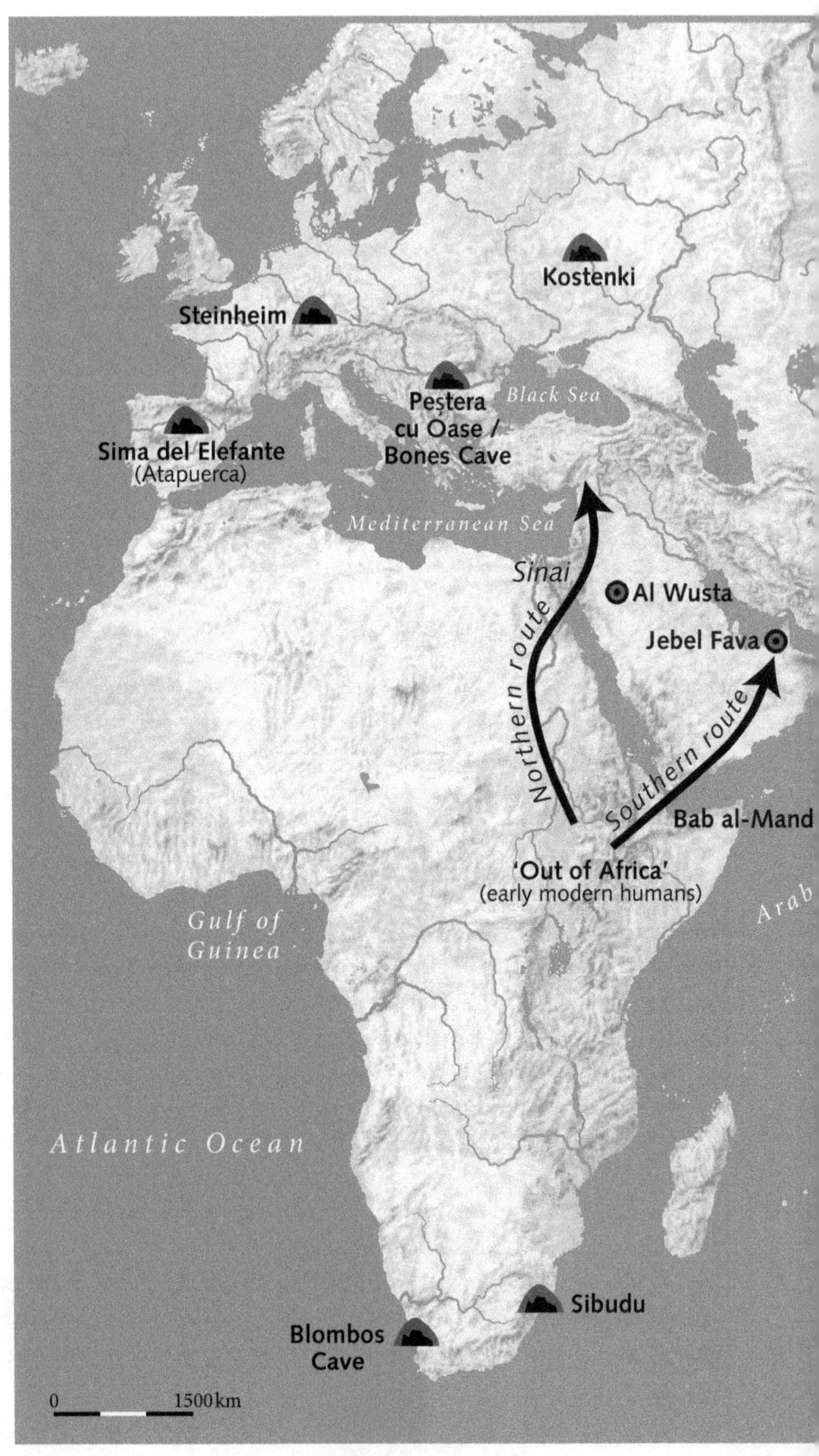

Figure 13 Apocalypse.

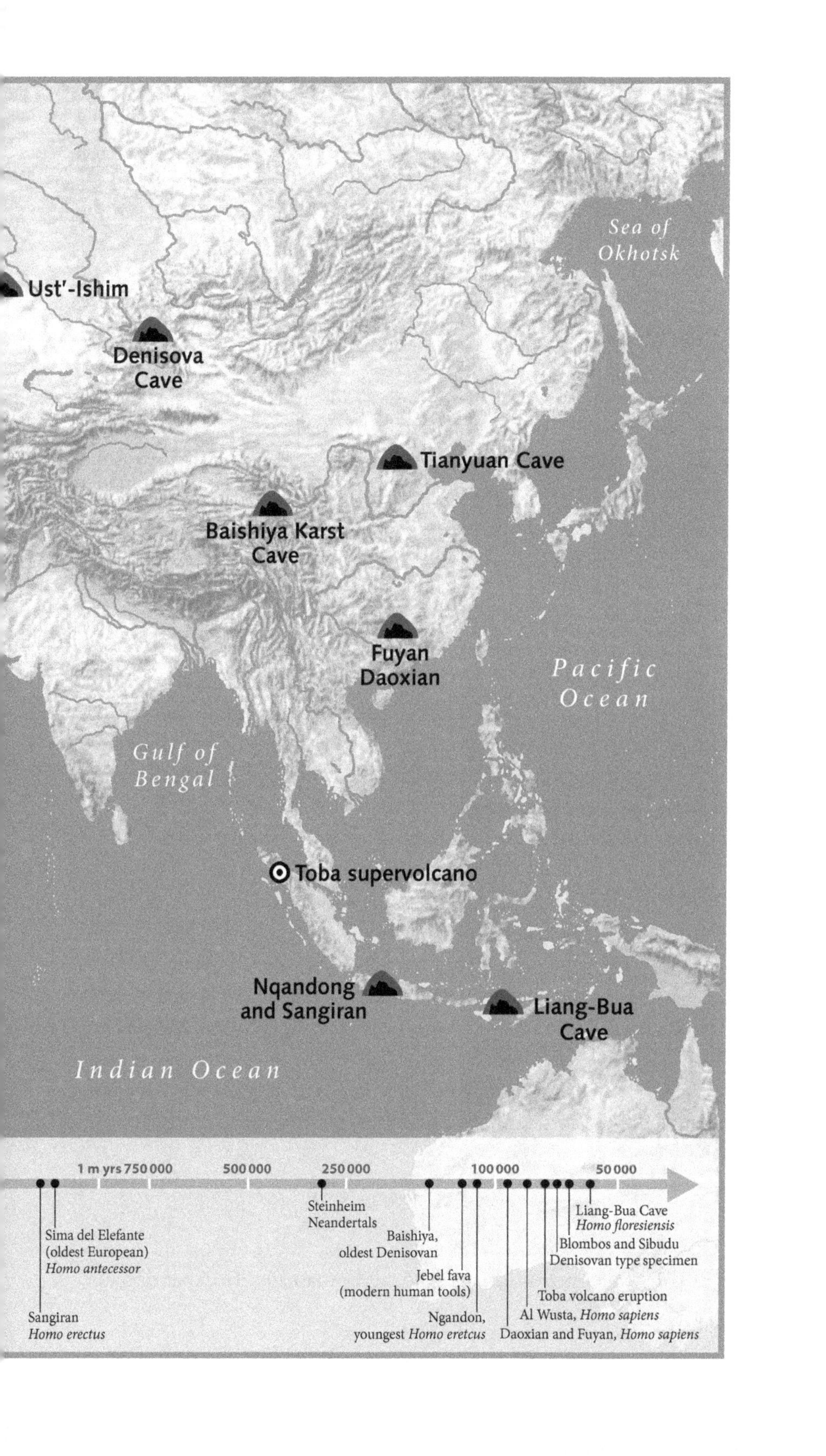

Sea of Okhotsk
Ust'-Ishim
Denisova Cave
Tianyuan Cave
Baishiya Karst Cave
Fuyan Daoxian
Pacific Ocean
Gulf of Bengal
Toba supervolcano
Nqandong and Sangiran
Liang-Bua Cave
Indian Ocean
1 m yrs
750 000
500 000
250 000
100 000
50 000
Sangiran *Homo erectus*
Sima del Elefante (oldest European) *Homo antecessor*
Steinheim Neandertals
Baishiya, oldest Denisovan
Jebel fava (modern human tools)
Ngandon, youngest *Homo eretcus*
Daoxian and Fuyan, *Homo sapiens*
Al Wusta, *Homo sapiens*
Toba volcano eruption
Denisovan type specimen
Blombos and Sibudu
Liang-Bua Cave *Homo floresiensis*

accessible point of Asia at least 1.5 million years ago; this point was Java in present-day Indonesia. It was still joined to the mainland at the time, as a result of the lower sea level. Given the not entirely conclusive dating of Java Man, it is also possible that this population arrived here soon after migrating from Africa. How long *Homo erectus* managed to survive in Eurasia is not clear, but the species probably still roamed through these parts 100,000 years ago and is therefore unlikely to have encountered our ancestors.

The older the archaeological finds, the more care is needed in their interpretation. In the case of bones that are hundreds of thousands of years old and usually preserved only in fragments, it is not always possible to distinguish clearly between *Homo erectus* and *Homo sapiens*, and the encyclopaedia entries for many sites often sound more definitive than is warranted by the data, in terms of both age and assignment to a particular type or species of human. The basic outline of the modern human was there in the upright walking *Homo erectus*, while the transition to *Homo sapiens* found its main outward expression in the altered shape of the skull, with its increasingly high forehead and head and diminishing brow ridge.

As any swimming pool user will confirm, the spectrum of the characteristics of *Homo sapiens* is enormous on its own. So one can imagine how difficult it is for archaeologists to reconstruct an entire body structure from ancient bone fragments and to assign it to the right species. This is further complicated by the existence of offshoots with archaic features but modern cultural techniques, such as *Homo naledi* in southern Africa. Some later finds that were classified as *Homo erectus* may actually have been examples of modern humans, especially as the term *sapiens* (wise) indicates a disposition that cannot be conclusively established from the size of the skull alone. Conversely, it is not always possible – particularly in a European context – to say for sure whether a bone find belongs to *Homo erectus* or to an early form of Neandertal: the roughly twenty-five-year-old woman who lived 300,000 years ago in Steinheim an der Murr, in what is today Baden-Württemberg in Germany, illustrates this ambiguous category.

The blackout of 70,000 years ago

One constant can be observed, however: as soon as the opportunity arose and their physical constitution allowed, all human species expanded into

Eurasia. And modern humans took this step long before those who were to become our ancestors. To date, there have been around forty relevant finds, all of which point more or less clearly in the same direction. The first examples of *sapiens*, it seems, found their way to the territories we now call China, Vietnam, and Indonesia over 100,000 years ago; at least the oldest remains discovered in China so far – mainly teeth believed to have belonged to modern humans – date from this period. In terms of timescale, this is consistent with the finds from Israel, where we know from the evidence of Skuhl Cave and Qafzeh Cave that modern humans existed around 120,000 years ago. And it also tallies with the climatic changes of that era: during the exceptionally long and at times extremely hot Eemian warm period, the conditions in southern Eurasia would have been highly congenial to humans who came from Africa.

Still, if we go back 200,000 years, to the time when modern humans inhabited southern Greece and the Misliya Cave site in Israel, we find no evidence of them in China and Southeast Asia. During this period, while the slightly mildening Ice Age may have allowed relatively modest advances northwards, it was perhaps not enough for a major expansion to the east.

From that time until the Eemian, indomitable cold may have turned parts of Eurasia into a virtually human-free zone where only Denisovans and Neandertals could survive. Whether these hominins came into conflict with *Homo erectus* is not known. The latter certainly still lived in Europe at the same time as the first Neandertals, so it is possible that they fought or mated with each other but, since we have no DNA from *Homo erectus* to date, this can't be tested. The same goes for Denisovans, though in their case there is the further problem that we don't even have a rough idea of when they first lived in Asia, and hence whether they would have come into contact with *Homo erectus* at all.

One thing is certain, however: China and Southeast Asia were not a success story for the first *Homo sapiens*. We know this because, after a period of apparently continuous expansion, the trail of modern humans suddenly dries up here about 70,000 years ago. It is as if they had spontaneously decided to turn their back on the Far East, to return only much later.

One of the earliest signs of a re-emergence of *Homo sapiens* is Tianyuan Man, a type that lived in the region of present-day Beijing some 40,000 years ago. This was, beyond doubt, a modern human and, more importantly, one of our ancestors. In 2013, a team from the MPI EVA succeeded in extracting DNA from the thigh and shin bones of Tianyuan Man. This DNA revealed a relationship with present-day East Asians and Native Americans, whose forebears migrated during the Ice Age via the Bering Strait (then dry, now a narrow waterway). The Tianyuan fossil is further proof of the rate at which our ancestors explored the world – and, again, raises the question of why modern humans seem not to have been present in East Asia between 70,000 and 45,000 years ago. The search for an answer leads us to Indonesia and to a global cooling event that, while it had nothing to do with the Ice Age, nevertheless greatly amplified its effects, especially in East Asia. In short, there is much evidence to suggest that the early pioneers of modern humanity fell victim to what was probably the largest known natural disaster of the past 100,000 years.

An explosive mixture

It is surely no accident that, for no specific reason, more than a hundred years of archaeological research have failed to uncover any signs of modern humans from a period of 25,000 years. Such a gap is unlikely to be attributable to a series of unsuccessful archaeological excavations. Even less plausible is the notion that the population that lived in Asia at that time spontaneously decided to withdraw from the region en masse. Something else must have happened – presumably something dramatic and terrible. When there is an archaeological bk hole, a deadly plague should always be considered as a possible explanation. But such a pandemic would have to have spread among a population of hunter-gatherers who were dispersed across the whole of Asia and only had sporadic contact with one another, if any. While a destructive confrontation with Denisovans cannot be entirely ruled out, it is highly unlikely: nothing in the subsequent course of history points to such an overwhelming dominance of this human type. No – virtually all the signs point to another cause of the Asian blackout of around 70,000 years ago: the eruption of the Toba supervolcano on Sumatra.

During the Ice Age, Sumatra was not yet an island but, like Java and Borneo, belonged to the landmass of Sundaland, which was joined to the Asian mainland. Mount Toba is idyllically situated on the lake of the same name, which is itself surrounded by mountains. At 900 metres, it looks anything but scary to a modern observer unfamiliar with its past. Around 74,000 years ago, about two thirds of the mountain exploded, producing the crater that now holds the picturesque 90-kilometre-wide lake. According to geologists' calculations, the 50-kilometre-high cloud that formed after the explosion contained 2,800 cubic kilometres of ash and darkened the sky over large parts of Asia for years. It was the biggest volcanic eruption the world has seen in the past 2 million years. For comparison, the eruption of Eyjafjallajökull in Iceland, which disrupted European air travel for several weeks in 2010, produced 0.14 cubic kilometres of ash. The humans living in the region of Mount Toba 74,000 years ago were most likely buried instantly under the shower of ash and lava, or else died from lack of oxygen. But the effects of the disaster continued to be felt for much longer – probably decades, if the model calculations of the subsequent cooling period are to be believed. According to these, the first few years may have seen the global temperature drop by up to 17 degrees. Needless to say, the regions most affected were those in the direct vicinity of the apocalyptic scene.

As the vegetation disappeared, so did the animals, and with them the basis of the hunter-gatherers' existence. But that was probably only the beginning, as a climate simulation involving NASA scientists demonstrated in 2021. The calculations showed that the volcanic eruption may have halved the global ozone volume and torn a huge hole in the ozone layer above the tropics – an impact that the study likened to the aftereffects of a nuclear war. The humans inhabiting this formerly hospitable part of the world may have suffered sight damage from increased ultra-violet (UV) radiation, along with massive sunburn; according to scenario projections, within just a few years the population would have been largely decimated by skin cancer and as a result of UV-damaged DNA. Plant and animal DNA could also have been affected, thereby reducing whatever sources of food were left under the darkened sky – which blocked the light but not the UV rays. The humans who had made their home in the east and southeast of China probably experienced an apocalypse that saw them and their genes buried in the dust of world history.

It is the Denisovans who were likely initially to have come out on top in the race for dominance over Asia. They, too, would certainly have been affected by the cooling period, but were probably better adapted to cope with it. The only two bone finds containing Denisovan DNA to date are from present-day Tibet and from the similarly glacial Altai Mountains, where the Denisovan girl 'Denny' lived, aptly enough, 70,000 years ago. Her ancestors or descendants, who lived almost 6,000 kilometres away from the volcano as the crow flies, must have been at a sufficiently safe distance to have survived the great Asian extinction. That said, the victims may still have included Denisovans, as some of them may have spread as far as Southeast Asia too: for them, the success of their kin in the north will of course have been small consolation. For the Denisovan population as a whole, however, the ability to survive at colder latitudes may have been what gave them their decisive evolutionary advantage over the modern humans of Asia. At least for a time.

In the case of *Homo floresiensis*, a species that probably lived at that time in the immediate vicinity of Mount Toba on what is now known as the island of Flores, we come up against the familiar challenges of archaeological dating: according to the calculated age of the few bones found in the area, this archaic human – often nicknamed 'Hobbit', chiefly because of its short stature – lived there up until 60,000 years ago. This could indicate a hardiness capable of withstanding the volcano – or, somewhat more probably, a certain range or margin of error in the dating: after all, *Homo floresiensis* was estimated to be 18,000 years old on its discovery in 2004, but subsequently turned out to be much older. Since the Mount Toba disaster itself can be dated only with approximation, a link between the extinction of the Hobbits and the eruption cannot be proved – but cannot be ruled out either. At all events, these mysterious Floresians, who somehow or other disappeared off the face of the earth before our ancestors' arrival, are worth a closer look, and we will come back to them later.

Unimaginable though the Mount Toba disaster is for people today, its impact shouldn't be overestimated. That it wiped out the majority of East Asians at that time seems plausible, as does the global cooling event. But the effects on humans in Africa must have been relatively limited, and a comparable mass extinction is highly unlikely to have happened there. The same cannot be said of southern Europe, however: there the conditions would have been just about mild enough to sustain African migrants 70,000 years ago, but even a slight drop in temperature could have spelt the end for them – unlike for Neandertals, who were so tough in this respect.

By contrast, the Toba disaster theory, which was advanced more than thirty years ago and remains highly controversial, is based on the hypothesis of a global catastrophe that brought the whole of humanity to the brink of destruction. Subscribers to this theory point for example to the genetic bottleneck through which our ancestors in Africa passed: the fact that all present-day humans can be traced back to a small group of Africans, they argue, can be explained by the prior decimation of the population caused by the direct and indirect consequences of the volcanic eruption. Mind you, this could be a circular argument, being as hard to prove as it is to disprove.

There is, after all, no doubt that such a bottleneck exists in the gene pool of present-day non-Africans – but all this really proves is that they are descended from a small population of origin. The fact that this bottleneck phase began after the eruption doesn't mean that it was necessarily caused by it. Moreover, there is no observable bottleneck effect in the genes of present-day Africans, yet it is clearly evident in a variety of other mammal populations, such as East Asian rats for example. Both these findings suggest an Asia-wide blackout after the eruption, but not a global one. This would make Mount Toba the end point of a first wave of modern human migration to eastern Eurasia, but by no means the beginning of the one that went on to displace all other human types on the planet.

North Africa, valley of death

The Mount Toba disaster falls within a period of roughly 20,000 years that is quite mysterious for archaeogeneticists, in that it is still impossible to say exactly what happened during these years – and, above all, where. The beginning and end points of this enigma can be determined unequivocally by genetic analyses. On the one hand, we know that the common ancestors of Africans and non-Africans parted company about 70,000 years ago, obviously in Africa, and probably somewhere in the east or southeast of the continent. The last common ancestors of humans outside Africa, on the other hand, lived some 50,000 years ago. Where the population that gave rise to the migrants was based in the intervening

20,000-year period is, however, completely unknown. Given that during this phase they interbred with Neandertals, whose genes are present in all non-Africans today, it must have been somewhere in the southern portion of the Neandertal domain – and in one of the many regions where archaeological traces of modern humans from that time have been found.

This leaves us looking at a very large area, which stretches from Egypt to the west of the Indian subcontinent. A glance at the atlas invites an obvious conclusion: if a group of Africans embarked on a northbound journey about 70,000 years ago, the most likely route would have taken them via present-day Egypt and the Sinai Peninsula – then as now, the only land connection between Africa and Eurasia. Since there were Neandertals living in Israel, the interbreeding could have taken place in the Middle East, after the Africans' arrival via this northern route.

When you look at the atlas, this theory seems irresistible; and it still remains a legitimate one. It is nonetheless contradicted by the latest climate models, which show that for the past 130,000 years almost the entire northern route was completely impassable to humans and that anyone who attempted it would have perished in the desert sands. According to these new estimates, another route appears much more likely, namely by sea. Our ancestors could have taken the southern route, across the Bab al-Mandab Strait, through which the Red Sea – the sea that separates East Africa from the Arabian Peninsula – flows into the Indian Ocean. There were only a few possible windows for crossing this strait, but they happen to coincide almost exactly with the genetic data on the migration of modern humans to Eurasia. Once again, we can see how inextricably the story of our expansion is bound up with the climate, and above all with its changing phases.

In 2020, a British team from Cambridge University published an extensive study that reconstructs the annual rainfall volumes in East and North Africa for the past 300,000 years – something that, like so many scientific achievements, was made possible only by the incredible processing power of the relevant computers. Just about every known factor that can be used to draw conclusions about climate and rainfall was fed into the model – be it ice core drillings in Greenland, global prehistoric water levels, or sediment deposits in bodies of water.

Using these proxy data, as they are called, a climate model was produced for the entire world, and hence also for Ice Age Africa. In order to determine which regions may have been inhabited by humans at that time, the authors of the study set the absolute minimum rainfall that would have been necessary for survival at 90 millimetres per year. Some places in East Asia occasionally get that much rain in a single day, and even America's Death Valley – which sustains what little life there is only because it catches the snowmelt from the Sierra Nevada – has a higher average precipitation.

Personnel shortages hamper evolution

As the calculations show, precipitation in Upper Egypt has been well below the 90-millimetre mark for most of the time during the past 300,000 years. Nor was the Nile – which rises in Central Africa and would later become the lifeblood of the Egyptian civilization – anything like a suitable corridor for early human travel to the north. Unlike its northern delta and lower reaches, the Nile is – and was – completely inaccessible, especially in its upper reaches in present-day Ethiopia, since in this area it flows through gorges that are sometimes one and a half kilometres deep. Although the banks of the river would have been reachable in the adjacent Sahara, there were no fertile valleys there capable of sustaining hunter-gatherers. These disastrous navigational conditions appear to have prevailed for almost the whole period during which modern humans developed in Africa – with one exception: about 130,000 years ago, during the Eemian Interglacial, the Sahara turned green when the North African climate became a lot wetter.

Judging by these climate models, it cannot have been via the northern route that the ancestors of all non-Africans dispersed; to this extent, Africa seems to have been cut off entirely from the rest of the world. The currents in the 14-kilometre-wide Strait of Gibraltar would have been far too strong for early humans, nor are there any archaeological or genetic clues to a successful migration via the Iberian Peninsula in prehistoric times. Then again, the other strait looks equally improbable as soon as you glance at the map: Bab al-Mandab, between present-day Yemen and Djibouti, has much stronger currents than the Strait of Gibraltar and is also twice as wide at its narrowest point, measuring 27 kilometres across. That's comparable with the length of the Fehmarn Belt between Germany and Denmark: the first time a person swam across this stretch of water was in 1939. So, strictly speaking, the chances that a number of humans from Africa made it to the Arabian Peninsula via the southern route 70,000 years ago would appear to be zero.[1] Strictly speaking.

According to the latest climate models, the drop in the sea level during particularly cold periods of the Ice Age caused the strait to shrink persistently, in some places to a width of just 5 kilometres – a distance that would surely have been manageable even for early humans, especially given the reduced water volume and hence much weaker currents. This in itself would not have been enough to warrant a passage from Africa to Arabia, however: climatic conditions conducive to life on both sides of the channel would have had to prevail. The models suggest that all the prerequisites were there – humans inhabiting the Horn of Africa, a sufficiently narrow strait, and fertile conditions in the north of the Arabian Peninsula – albeit on rare occasions: the first time 250,000 years ago, then again 130,000 years ago, and finally 65,000 years ago. In the latter period – importantly, as it would turn out – the strait remained passable for an incredibly long time, and certainly much longer than during the previous sea level reductions.

The climate models tally almost exactly with the archaeological finds that testify to modern humans' first attempts to migrate to Europe and Asia. The ancestors of the individual who inhabited Apidima Cave in southern Greece 220,000 years ago may have reached Arabia a quarter of a million years ago, on rafts or tree trunks lashed together; the same goes for the forebears of the Misliya people discovered in Israel 180,000 years ago. As for the 120,000-year-old occupants of the northern Israeli caves Skhul and Qafzeh, they could have migrated there via either the southern or the northern route, the latter being passable at that time. The common ancestors of all non-Africans, on the other hand, would have had to cross the water again when they began arriving in Eurasia 65,000 years ago.

But why did it take at least three attempts before our ancestors succeeded in spreading out from Africa around the globe? The interval between the first appearance of modern humans in the Greek Peloponnese and the eruption on Mount Toba was nearly 140,000 years – time enough to conquer Eurasia. This natural disaster alone, then, can hardly be to blame for their repeated failure. It took another one to cap it off: the persistently rising sea level, which cut the migrants off from their original population, isolating them on the Arabian Peninsula. The resulting loss of genetic exchange may have been the final nail in the coffin of modern humans' early attempts at expansion. On their own, they didn't stand a chance – unlike their brothers and sisters in Africa, who achieved new cultural and technical breakthroughs during this period.

Remember: not just present-day Eurasians but all Africans go back to the population that formed in the melting pot of Africa starting some 300,000 years ago. It was there that *Homo sapiens* evolved and, with it, the complex culture of modern humanity, which subsequently spread in all directions, including on the neighbouring continent to the north. From South African caves, for example, we have archaeological evidence of the use of ochre 100,000 years ago, in the form of geometric patterns carved into the material; it was probably also used as a cosmetic (Figures 14 and 15). About 70,000 years ago, the occupants of Sibudu Cave – also in South Africa – appear to have lined their beds with leaves, which they burned at regular intervals to rid their sleeping quarters of vermin. Signs of the first bows and arrows in human history were also found in this cave, and were dated to a period around 65,000 years ago. The small stone blades of roughly the same age that our ancestors managed to fashion are far more complex than anything we have seen made by Neandertals, who generally had the necessary materials but not, it seems, the relevant know-how.

All this was no accident, but undoubtedly had a lot to do with a surplus of labour among our African ancestors. This is because, for all the new technologies and skills that humans developed, they needed a sufficiently large population to allow talented individuals time outside the business of hunting and gathering, so that they could still share in its fruits, even if their own activities were not related to survival.

In this respect, a fertile place like Africa offered an ideal environment. True, there was no population explosion in the days of hunter-gatherers; that took another few tens of thousands of years, until the development of agriculture. But the continuous improvement of hunting techniques must surely have increased the availability of food, and hence the size of the population. Thus humans created both the conditions and the

Figure 14 The oldest evidence of cosmetics. Image provided courtesy of Christopher Henshilwood. Photo credit: Christopher Henshilwood.

The oldest evidence of cosmetics to date was found in the South African Blombos Cave. Early humans ground ochre to a powder in a stone bowl and presumably rubbed it into their skin. This earth pigment was also used for carving geometric patterns.

need for new hunting grounds and territory; whether they were simply following their prey or the human impulse for discovery was already at play, no one knows. But set off they did.

Burning bridges

Whatever drove the first *Homo sapiens* in the direction of Greece more than 200,000 years ago, their expedition was an ill-fated one. In the event, the rainfall and the climate on the Arabian Peninsula soon changed the prevailing living conditions in such a way that it became impossible to survive there, let alone progress any further northwards. The few thousands of years during which transit between Africa and Arabia was possible were probably insufficient to allow the expansion of a strong, resilient population capable of holding out against the Neandertals and the northern cold. The changing climate and consequent dwindling of

Figure 15 Ancient rock paintings by the San people © picture alliance / blickwinkel / McPHOTO

Ancient rock paintings by the San people – ancestors of modern-day Indigenous South Africans – also point to a very early artistic sensibility in the modern human, whose first 'studios' were probably here in Africa.

rainfall in Arabia represented a worst-case scenario for humanity, both in terms of population genetics and in terms of the progress of civilization.

From now on, the settlers were cut off from the cultural and technical developments of their relatives in Africa. Not only that, but their new home offered extremely poor conditions for producing the surplus of labour and creativity they needed for their own innovations. For, without new settlers, a relatively small original population of perhaps a few hundred will sooner or later lose the ability to 'renew' itself genetically. This process is crucial in a healthy population; otherwise genes liable to produce a lower level of fitness and other evolutionary disadvantages will not be filtered out.

Things were very different during the second known wave. This time, around 130,000 years ago, we know of several modern humans who made it not just to present-day Israel but probably as far as Southeast Asia. Possible routes include the strait between Africa and the Arabian

Peninsula that was narrowing once again, or else present-day Egypt, which may have served as a gateway to the north at that time. When we talk of an expansion via the northern route, we should not imagine, of course, a trail of bold early humans having decided one day to follow the North Star and build a new future elsewhere. The more likely explanation is that they pushed further and further into the fertile regions, over generations, until they eventually reached the Sinai Peninsula. As for the south of the Arabian Peninsula, this may have been reached at the same time via the strait. Whatever the circumstances, recent archaeological finds suggest that humans were already living on the Arabian Peninsula at that time. In 2018, a team from the MPI in Jena found an approximately 90,000-year-old finger bone from a modern human in Saudi Arabia.

This expansion, then, was extremely successful, as the protagonists advanced not just into the borderlands of Neandertal territory but apparently right to the furthest tip of Asia. For tens of thousands of years, humans were on a path to success similar to the one pursued later by our direct ancestors – until a promising future was suddenly buried in the debris of Mount Toba. For a long time afterwards, modern humans were conspicuous by their absence in Southeast Asia, while in eastern Africa the last configuration was taking shape – the one that would finally succeed in breaking through.

The price of isolation

The effect of small, isolated populations on evolutionary development can best be illustrated with a fictional island. If ten people arrive on this island and one of them exhibits a genetically related lack of fitness but an advantage for some reason in the search for a mate, then half of the second generation of islanders, and perhaps as much as 75 per cent of the third, may inherit the same disadvantage. If no newcomers with more favourable genes join this population, its chances of evolutionary success will lessen. This phenomenon, known in population genetics as 'drift', often rests on pure coincidence; it increases during genetic bottlenecks and perpetuates itself for generations thereafter. It is not always the case that harmful genes will prevail, but the likelihood of more favourable ones coming along is small.

This situation is compounded by another effect. In the case of closely related parents – as is inevitably common in isolated and very small populations – there is a higher probability that genetic mutations detrimental to the fitness of the offspring will combine with each other. This disadvantage was particularly pronounced among the last Neandertals and Denisovans, for example, who often practised outright incest. But when the first modern humans came to Europe, the boot would have been on the other foot: at that time they were probably the ones who, genetically, suffered most from the isolation of the north and were thus at an evolutionary disadvantage vis-à-vis the Neandertals.

Ghost DNA

How many humans – if any – continued to inhabit the region that stretched from the Middle East to India after the volcanic eruption, we don't know. All we know is that their genetic trail disappeared. Perhaps, when access to Arabia from the Horn of Africa opened up again 65,000 years ago and our direct ancestors embarked on their journey, they came across their relatives, who had retreated after losing out to the Neandertals in the north of the peninsula. Or perhaps that population had disappeared altogether by then. Either way, the new arrivals fared very differently from their predecessors: climate models suggest that the strait remained open until 30,000 years ago – more than enough time to build up a genetically robust population on the Arabian Peninsula.

That said, this one was not exempt from false starts either, as we know from numerous bone finds – beginning with Zlatý kůň and the 45,000-year-old bones of the Bacho Kiro people, whose DNA hit a genetic dead end, just like those of Ust'Ishim Man and Oase Man 44,000 and 40,000 years ago respectively. It is not until Tianyuan Man, 40,000 years ago in modern-day China, and Kostenki Man, 38,000 years ago in modern-day Russia, that we find the successful establishment of a genetic lineage that extends to the present. Behind them lay almost 10,000 years of fruitless assaults on the fortress of Eurasia, which – particularly in the west of the continent – was in the grip of the Neandertals.

All these migrants shared an element of Neandertal ancestry. Yet, paradoxically, that element is smaller today in Europeans and Middle Easterners than it is in the rest of the non-African world. The reason for this could be what we call the Basal Eurasians: a non-localizable 'ghost population' whose genes we are aware of, but only because they were found to exist in later populations: to date, no Basal Eurasian bones have been found, let alone direct DNA evidence.[2] There is now much to suggest that this group lived in the Middle East too, some 65,000 years ago. But, unlike their fellow humans, Basal Eurasians don't appear to have mixed with the Neandertals; at least there are no traces of these prehistoric humans in the Basal European DNA components discovered so far. Which means that they must have split off from the group of African migrants before the latter interbred with the Neandertals.[3]

There is no Basal Eurasian DNA in the genomes of the Ice Age hunter-gatherers of Eurasia, Australia, or America either, but it is present in the genomes of all humans indigenous to the Middle East and North Africa from around 15,000 to 8,000 years ago. The Natufians,[4] for example, who lived in present-day Israel and Jordan 14,000 years ago and went on to become the world's first arable farmers, carried roughly 40 per cent of Basal Eurasian DNA, as did the inhabitants of present-day Iran. The remaining DNA of the humans in this region came from the population that had previously mixed with the Neandertals. Put simply, the ancestors of the first arable farmers in the Middle East must have interbred, and pretty copiously at that, with Basal Eurasians – in contrast with all other humans who subsequently migrated to Europe and Asia. This is perhaps a rather complex genetic background, which, although it doesn't provide conclusive answers, at least allows an educated guess.

To this day, we still have no clear picture of where and when the Basal Eurasians lived. But if their DNA was so dominant in the people who later settled in the Middle East, yet nowhere else, then the Basal Eurasians must have come from the same region. Judging from climate models and all the available genetic data, one scenario thus appears to be highly plausible: that the Basal Eurasians were the people who migrated from East Africa to the Arabian Peninsula beginning 65,000 years ago. They could also be described as Basal non-Africans. This population then split again, so that a larger group could have migrated north and bred with the Neandertals there. Or the interbreeding occurred throughout the Arabian Peninsula but the Basal Eurasians migrated beforehand, to a region where there were no Neandertals.

One way or another, there must have been a natural barrier between the two populations for tens of thousands of years, otherwise there would have been recurrent genetic exchanges and we wouldn't have this ghost DNA today. Perhaps the Arabian Desert suddenly turned from a green corridor into a no longer navigable death zone, of the kind found in present-day Saudi Arabia. Or, equally conceivably, the first Basal Eurasians crossed the Zagros Mountains of Iran, and then the way back became blocked as the Ice Age grew colder. The Lut Desert too, in present-day Iran and Afghanistan, could have acted like a door that was opened for a time, only to be slammed shut again.

While Basal Eurasians were cut off from the common ancestors of all Eurasians, the others began to conquer the world. Eventually Basal Eurasians did mix with their neighbours – probably when the barrier became fertile or melted. This may have been 30,000, or perhaps 20,000 years ago. All we know is that, some 15,000 years ago, the genome of Middle Easterners derived from both populations, in roughly equal measure. Since Basal Eurasians carried no Neandertal DNA, they must be the ones who also diluted the Neandertal element in the first arable farmers of the Middle East.

Basal Eurasian DNA then finally came to Europe 8,000 years ago, when Anatolian farmers displaced the continent's hunter-gatherers. This genetic shift produced a similar reduction in Neandertal ancestry – something that remains unchanged to this day. In Asia, which the Anatolians failed to reach, it was another story: there the Neandertal element remained constant, and hence higher than in Europe. And the same applied to the Indigenous populations of America and Australia. The separation process that occurred soon after our ancestors' migration out of Africa is etched in our genes to this day. Thus present-day Europeans have an average of 2 per cent Neandertal DNA, while East Asians, Native Americans, and Australians have around half a percentage point more.

From dream to illusion

Once modern humans had matured in Africa into the *sapiens* type and consolidated their population on the Arabian Peninsula, the way back to Africa remained closed for a long time. According to recent climate models, some 30,000 years ago rainfall on either side of the strait appears to have diminished to a point where this inhospitable region once again ceased to serve as a corridor between East Africa and the Arabian Peninsula. Africans were cut off for ages from the rest of the world, until, about 10,000 years ago, the Sahara turned green again and the northern corridor became passable once more. A similar fate befell the Americans and Australians, who were able to reach their new home during the Ice Age thanks to falling sea levels, but remained isolated from the original Eurasian population for thousands of years when the planet warmed up again.

Millennia were to pass, however, before our ancestors finally managed to do what countless humans before them had failed to accomplish: conquer the land of Neandertals and Denisovans, of wolves and hyenas, of ice and steppes. By the end of that journey, human culture had triumphed, for the first time in evolutionary history, over the logic of biology. Having withstood everything that the Ice Age, famine, and the cruel hand of nature could throw at them, Neandertals and other early humans ultimately failed to survive the arrival of civilization. But before they bowed out, our ancestors managed to filch a few handy genes from them. Genes that would help us, for example, to reach the roof of the world.

5

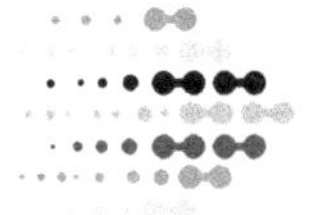

A Clean Sweep

We turn our attention to the Neandertals and the Denisovans one last time. They both occupy their own niche, until we arrive on the scene. We emancipate ourselves from nature's constraints and subjugate it to our will. Large animals are consumed until there are no more left.

Limited scope for snap dates

To this day, the bottleneck effect that migration to the Middle East exerted on our African ancestors can be read in the genes of non-Africans. While the current world population is descended from a group of around 10,000 people, in the case of non-Africans that group is only half as big. The migrants represented only a part of the gene pool; consequently, all DNA in existence today outside sub-Saharan Africa is of African origin, apart from the Neandertal and Denisovan elements. The ancestors of the future Eurasians numbered around 5,000 individuals: even if modest by modern standards, from the perspective of those days such a group was certainly big enough to constitute a dense population in the Middle East.

Although the genomes of the Neandertals also point to a maximum long-term population size of around 5,000, this population was spread across the whole of Europe and half of Asia. Indeed, according to these data, the number of Neandertals even fell at times – whether because of food shortages or because of periods of extreme frost – to as few as 500 individuals, which brought them close to extinction. The assumption that these were the phases in which the Neandertals showed a stronger tendency towards cannibalism is not implausible. The theory of an

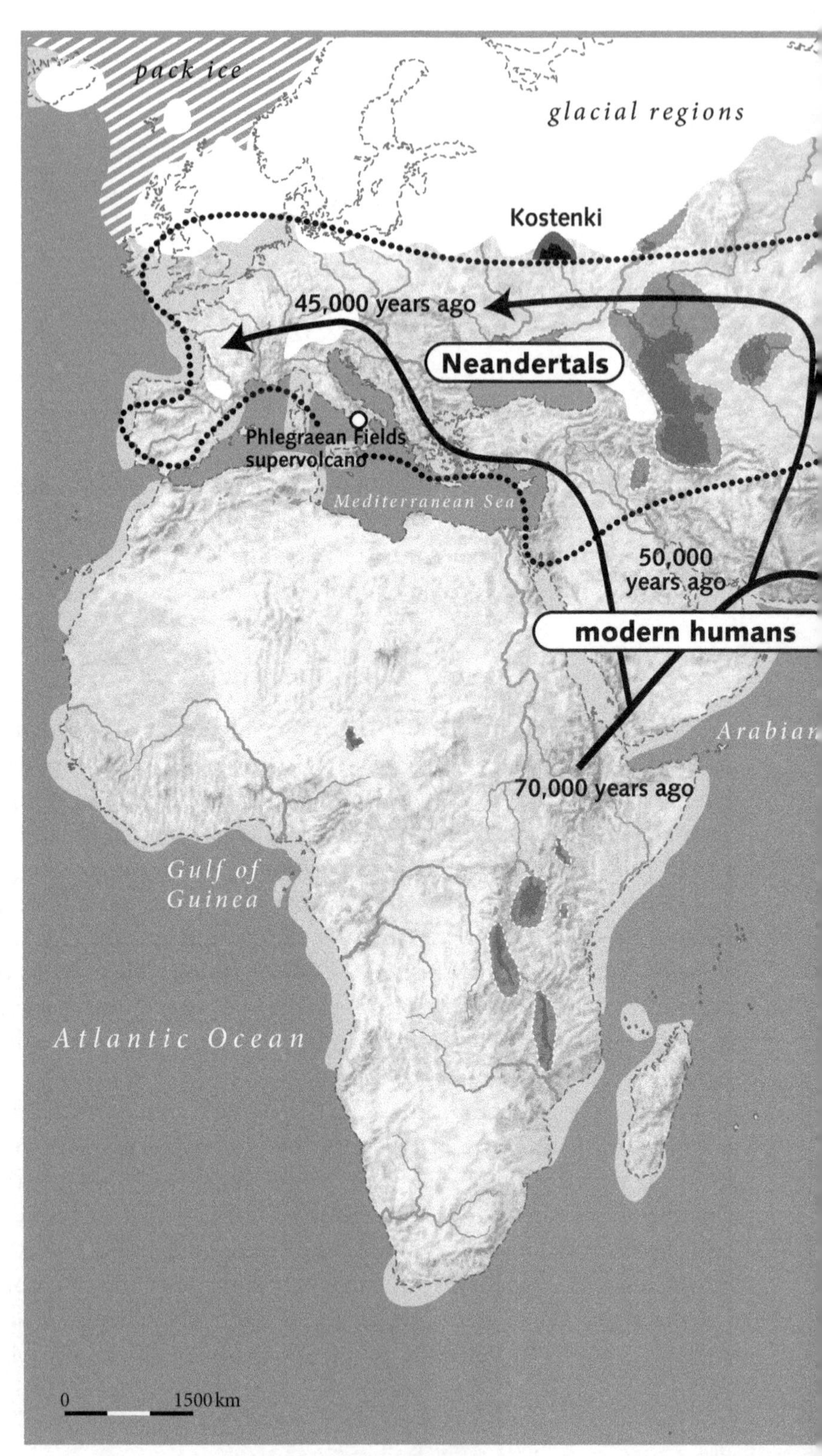

Figure 16 A clean sweep.

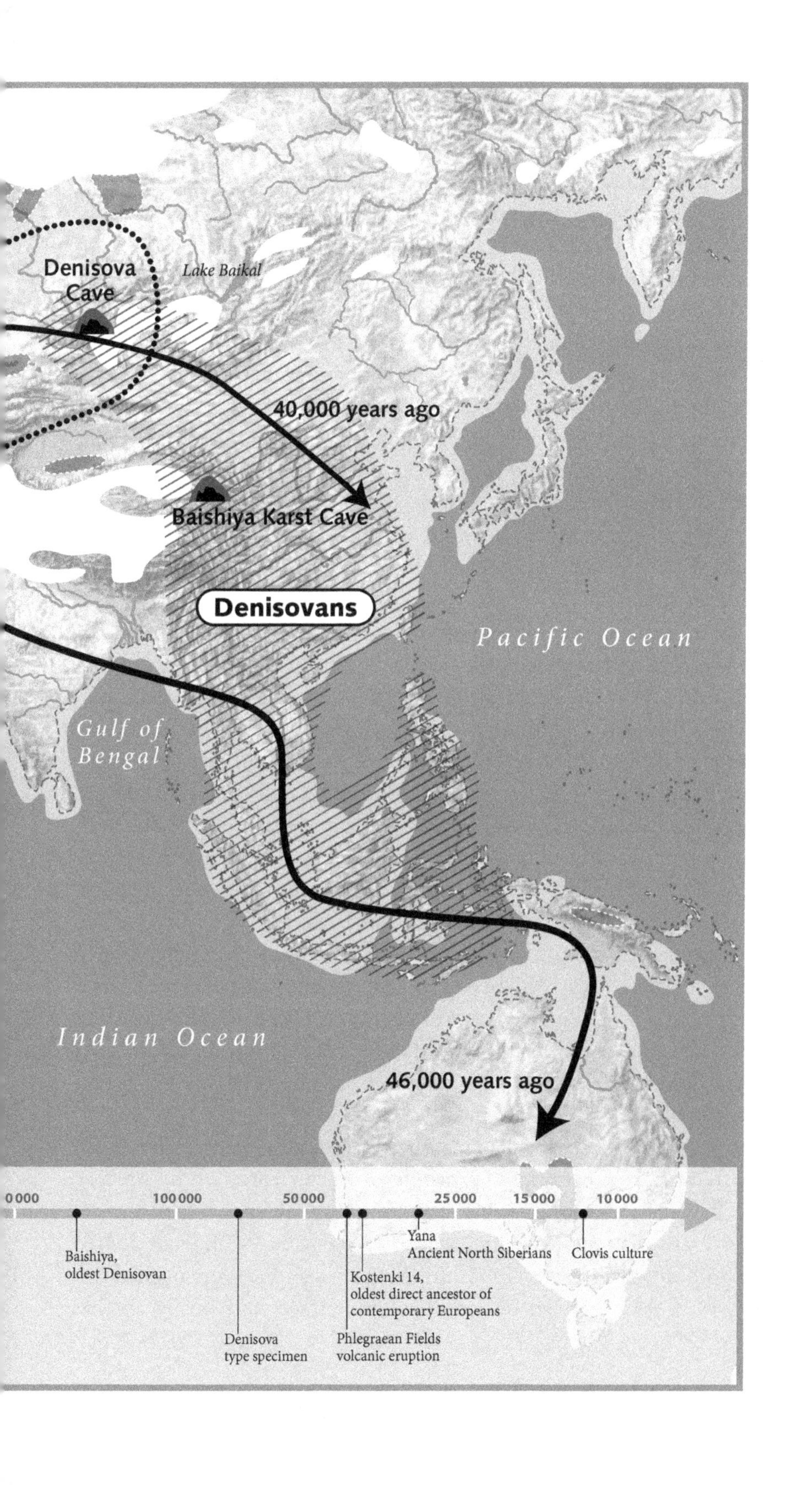
Denisova Cave
Lake Baikal
40,000 years ago
Baishiya Karst Cave
Denisovans
Pacific Ocean
Gulf of Bengal
Indian Ocean
46,000 years ago
0000
100000
50000
25000
15000
10000
Baishiya,
oldest Denisovan
Denisova
type specimen
Phlegraean Fields
volcanic eruption
Kostenki 14,
oldest direct ancestor of
contemporary Europeans
Yana
Ancient North Siberians
Clovis culture

excessively small population also fits with the accumulation of potentially harmful mutations in the Neandertal genomes analysed to date, which suggests that interbreeding often took place between very close relatives.

A relevant proportion of the already very few early humans could therefore have been carrying quite some genetic baggage by the time our ancestors came into contact with them. While the folk in the Middle East had a steady influx of new blood from Africa, and were thus able to have their genes mutually exchanged and repaired within a closely coexisting population, this was generally not the case with the Neandertals: when no more than a few thousand individuals are spread out all the way from Israel to the Rhineland and from Spain to Siberia (to use contemporary names), spontaneous sexual encounters are not exactly common. And, even when they occur, the partners may well share the same (great-) grandparents.

In short, the chances of bringing resilient children into the world and thus building a healthy population size were clearly much higher among our ancestors than among the Neandertals. On the other hand, the latter had an undeniable home advantage in Europe. They had lived for hundreds of thousands of years in the Ice Age mammoth steppe, which stretched all the way from France to Central Asia, and they knew it like no one else. They had developed an astonishing expertise in mammoth and other big game hunting, and above all a fearlessness to match. Given their distinctly primitive spears, it must have been no mean feat to approach these metre-high animals and slay them from a trunk's length away. It takes time to build up that kind of experience over generations, just as it does to learn which plants, berries, and fungi go well with mammoth and which ones are more likely to prove one's last meal on earth. When modern humans set off for the north, with its completely different flora and fauna, they had none of this vital local knowledge at their disposal.

This may have been another reason why so many of them failed in their attempts at expansion: the people of Bacho Kiro, whose success in the Balkans was short-lived; Zlatý kůň, who was possibly eaten by hyenas; Oase Man in Romania, whose 10 per cent Neandertal DNA didn't do him any favours; Ust'-Ishim Man, whose ancestors pushed far into Central Asia, only to disappear without trace; and finally all the modern humans whose excursions to the north left no genes behind, and not even any bones.

The Neandertals in the west and the Denisovans in the east of Eurasia had occupied their respective niches, and their existence reflected the natural course of life. They hunted mammoths and other animals of the steppe within their means and, since these were fairly limited, the relative numbers of humans and animals remained in equilibrium. The manpower for technical innovations wasn't there, as everyone had to devote all their energy to day-to-day survival, not to things like bows and arrows, micro-blades, or spear throwers.

Conditions were very different in the much milder Middle East. During the Ice Age this region offered a rich flora and fauna, just like other large parts of Africa, thus paving the way for population growth and, with it, an evolving capacity for innovation. Over countless generations, humans increased their knowledge of the natural world around them and gradually developed the technology they needed to survive in colder climates. Having already discovered fire, they now learned to make clothes from animal furs using needles and sinews. Soon our ancestors also acquired the core competence of the Neandertals: hunting megafauna. Yet despite all this they still couldn't compete with Eurasia's prehistoric humans. The fact that they succeeded in the end could have a little to do with their adoption of so many genetic traits from the Neandertals. It was like slipping into their skin.

Chain-smoking Neandertals?

One of the first genetic variants our ancestors were found to have acquired by interbreeding with the Neandertals was indeed the one responsible for a thicker skin. The benefit of such a genetic peculiarity in the cold north is obvious, and it also seems to have given our ancestors a clear selective advantage, as 70 to 80 per cent of all Eurasians alive today have the Neandertal variant in the part of their genome that governs the production of keratin. Keratin is a protein involved in building not just skin cells but hair and fingernail cells too. Increased production of this protein makes for tougher and also thicker skin, thereby reducing heat loss.

The same effect may have resulted from the Neandertals' pigmentation genes. But the findings regarding this genetic variant are less clear-cut than in the case of skin thickness, since here we are dealing with a whole class of genes whose exact functions and interactions are hard to distinguish. We don't even know whether the Neandertals' pigmentation genes are responsible for a darker or a paler skin: the complexion of contemporary humans who carry this variant allows for either of these possibilities. All the same, it is of course more likely that modern humans acquired a paler skin from the Neandertals, given that less pigmentation would have enabled them to produce more vitamin D from the comparatively fewer sunlight hours in the north.

Neandertal pigmentation genes have been found in contemporary Britons with both hues of skin – darker than average and lighter than average. These genetic variants are further evidence that the external properties of genes – which constitute the phenotype – cannot always be clearly determined, not even in the case of obvious physical traits such as skin colour. The genome is far more complex than even the best computer in the world, because it doesn't consist of binary codes but of variations and combinations of the billions of items of information stored in base pairs.

An even clearer illustration is that of the 'smoker gene'. Today this genetic variant inherited from the Neandertals is found with above-average frequency in heavy smokers, and is therefore associated with a higher potential for addiction to tobacco, or rather to its nicotine content. There are two things we can say with certainty here. First, there was no tobacco in Neandertal times. And, second, a predilection for one of the most dangerous neurotoxins and for the carcinogenic substances contained in tobacco smoke is not a survival advantage for any living creature and must, in theory, constitute an evolutionary disadvantage. Both these things show that, besides an affinity for tobacco, the genetic variant in question must produce at least one other trait that grants the carrier a 'selective advantage'. What that is, we don't know. All we know is that, had the Neandertals mastered the art of rolling cigarettes and growing tobacco, the air in their caves would have been pretty foul. Incidentally, the same genetic variant is not associated with other addictions.

Sensitive Americans

If chain-smoking Neandertals were alive today, their probably less than healthy respiratory tracts are likely to have been much more susceptible to the SARS-CoV-2 virus. There is, however, another reason why the pandemic would have hit them even harder than it did us – at least according to the comparative genome analyses conducted on Covid-19 patients in 2020. Right from the beginning of the pandemic, the Finnish genome research project FinnGen began retrieving, from the hundreds of thousands of genomes in their database, those of people who had contracted Covid-19 or died from it. Their aim was to identify DNA variants that increased the risk of severe illness and mortality – a prime example of the medical benefits, often direct, of genome research.

Having identified certain gene positions that were strikingly prevalent in patients who died from, or contracted, severe cases of Covid-19, FinnGen made these data available to the international research community, whereupon Neandertal researchers at Stockholm's Karolinska Institute and at the MPI EVA in Jena ran it through their computers. And, lo and behold, the Swedes identified a DNA fragment that consists of around 50,000 base pairs on chromosome 1 and that gives those with the Neandertal variant there a threefold risk of dying from Covid-19.

The genome segment in question, it seems, has a 'trans effect' on other genes that could have an important function in the event of a Covid-19 infection. One of these is the gene CCR9, which governs what we call a chemokine receptor (CCR) responsible for controlling a part of the body's immune response. At the same time, an adjacent gene interacts directly with the angiotensin-converting enzyme 2 (ACE2) receptor. This protein, which is involved in the transmission of signals within the body, provides the coronavirus with an entry point and is suspected of being the cause of the higher mortality rate in men, who tend to have a higher concentration of ACE2 in their cell membranes.

Another DNA segment that some people outside Africa have inherited from Neandertals appears, by contrast, to protect against severe cases of SARS-CoV-2. This segment, on chromosome 12, is responsible among other things for genes that play a key role in ribonucleic acid (RNA) virus infections.

Since Neandertals would definitely not have encountered SARS-CoV-2 viruses or, in all likelihood, any other coronaviruses, it may be that this gene expression had no advantage or disadvantage for them, but simply spread according to the laws of chance. Or, alternatively – and this is the far more likely explanation – it had another associated characteristic that increased the carrier's chances of survival or reproduction. The distribution of these Neandertal segments in present-day humans certainly provides some relevant clues. For one thing, this variant is almost non-existent in Europe, where it occurs only in 1–2 per cent of the population.[1] In Pakistan, on the other hand, it is present in almost half the population. This could have something to do with the high prevalence of cholera bacteria in that part of the world.

According to the latest clinical data, the effects of the toxin produced by cholera bacteria, which can cause deadly diarrhoea in humans, appear to be weakened by a receptor influenced by the CCR9 gene. It may be, then, that Neandertals with a more effective CCR9 gene enjoyed better protection from dying of cholera or other gastrointestinal infections – a selective advantage that, after the interbreeding event, became established among modern humans in parts of the world that were particularly afflicted by such diseases.

Another genetic characteristic of the Neandertals seems to have had special health benefits for modern humans in the cold north. In contemporary Europeans at least, there is an above-average incidence of a Neandertal gene segment that acts on one of the toll-like receptors (TLRs), which are in turn responsible for our innate immune system. It is probable, if not certain, that this was an adaptation to bacteria or viruses that were present in Eurasia at that time and were harmful to humans. Exactly which pathogen was involved is not known, however.

A further Neandertal gene segment in the contemporary global gene pool has remained something of an enigma so far, and it is not at all clear whether it spread by chance or perhaps offered some kind of advantage. One thing we can say is that this gene is probably not good news for carriers, as it causes very high sensitivity to pain. Judging by modern standards, Neandertals had it in spades: they exhibited a pain threshold roughly equivalent to that of an eight-year-old. We know this because the pain receptor encoded by this Neandertal gene was modelled in the laboratory and its sensitivity to electrical signals was measured.

The Neandertal pain gene has barely survived in Europeans, being carried by less than 1 per cent of the population. Among native Mexicans and South Americans, on the other hand, the relevant gene segment is present in almost one in two people. One explanation for this is that it became established after the colonization of America via the Bering Strait because it offered a selective advantage in those parts – be it through higher sensitivity to pain or through some other gene trait that we have yet to discover. Alternatively, it may have manifested itself in the bottleneck population that migrated from Asia to America – in which case its prevalence in present-day America would simply be a genetic coincidence. That said, the idea that the higher sensitivity to pain that modern humans inherited from the Neandertals could have offered an advantage when it came to discovering an unknown continent is not as bizarre as it might seem: it could protect people in regions unfamiliar to them, for example by triggering a warning signal to the body in response to poisonous plants and the like. But how this genetic characteristic might have benefited Neandertals, who knew their surrounding flora and fauna like the back of their hand, remains a mystery.

Champion childbearers

To attribute the demise of the Neandertals solely to the genetic superiority of the incomers would certainly be an overstatement. At the same time, genetic factors cannot be ignored if we want to understand why, in the days when a variety of hominins existed side by side, our ancestors felt persistently compelled to push ever further northwards, while the Neandertals for their part showed no ambitions to conquer the south. Or why, over a period spanning several hundreds of thousands of years, the Neandertals were incapable of adapting to living conditions that differed from those they were used to, while modern humans seem to have been able to switch from the African savannah to the Eurasian steppe effortlessly, in an evolutionary blink of an eye.

As is so often the case, there is no single answer to this question and, if one of the answers lies in our genes, we haven't found it yet. We see instead the key factor that obviously distinguished our ancestors from their doomed cousins: their complex human culture and associated ability to appropriate and shape nature. And, in interbreeding with the Neandertals, our ancestors appropriated this aspect of 'nature' too, quite literally by incorporating it – even if they did so unconsciously. What remained of the Neandertals in the end was something of considerable value to us: their niftiest genes. That's not to say, of course, that the Neandertals were merely gene donors for our ancestors: they too may have profited from the new gene pool that was flowing into Eurasia – though clearly not in a manner that allowed them to survive in the long term.

Paradoxically, the Neandertals were actually better equipped to survive as a population, at least as far as the physical constitution of women was concerned. As we know, the death of one's offspring at birth or in the first few years of life was par for the course right into the twentieth century, and this tragic fate had accompanied humanity since the beginning of its evolution. The Neandertals, however, seem to have been less afflicted in this respect than our ancestors. In 2020, a collaborative study by the Karolinksa Institute and the MPI EVA showed that one Neandertal genome segment is associated with a significantly lower rate of miscarriages and stillbirths in contemporary women. This is the single gene that encodes the progesterone receptor. In women with Neandertal DNA in the relevant genome segment, more of these receptors are produced. The hormone progesterone in turn causes a thickening of the uterine lining, and hence better blood supply to the embryo; for this reason, synthetic progesterone is sometimes used in fertility treatment or for women who have suffered repeated miscarriages.

It seems therefore that Neandertal women were generally more fertile as a result of having a progesterone production that was, on average, higher. This should have been a clear biological advantage in building a larger population. It may have been significantly offset, however, by the better chances of survival enjoyed by modern humans in the milder south. But when the latter reached the north and began to breed with the Neandertals, they picked up this selective advantage from them as well. According to a study based on British data, the relevant Neandertal gene segment is found in around 60 per cent of present-day European women.

Culture trumps biology

About 10,000 years after receiving the first visit from their southern cousins, 39,000 years ago at the latest, the Neandertals disappeared and ceased to exist as a human species, living on only as a marginal gene segment in us. Their displacement by our ancestors could have been fuelled by global warming. The latest climate models show that the glaciation previously assumed to have affected half the Eurasian continent 40,000 years ago was in fact confined to Scandinavian latitudes. After tens of thousands of years of extremely frosty conditions in Eurasia, our ancestors evidently took advantage of a mildening period to make their first forays.

It was these conditions that enabled them to establish in the Middle East, over a period of more than 20,000 years, a hunter-gatherer culture superior to the Neandertals'. The north now offered them space to continue and refine their project, while the Neandertals were presumably no longer able to oppose this progress – the product of millennia and favourable environmental conditions. The power balance shifted increasingly to the Neandertals' disadvantage until a tipping point must have been finally reached. The genetic adaptation to extreme environmental conditions that had guaranteed their survival in the past had become virtually worthless to them. That adaptability alone was no longer the engine of progress. Human culture trumped biology.

In the face of modern humans' advance into every corner of Eurasia, the Neandertals, who were mainly concentrated in the mammoth steppe, were soon left with nowhere to go. Already weakened by their long defensive struggle, they may have been finished off in the end by a natural disaster that occurred 39,000 years ago: the volcanic eruption of the Phlegraean Fields, close to Vesuvius. This eruption was not quite as dramatic as that of Mount Toba about 35,000 years before, but it struck at the heart of Neandertal territory. With clouds of ash that darkened the skies across Europe for years to come, the vegetation – and hence also the mammoth population – must have taken a dramatic hit. The time was ripe for those who could discover new sources of food, be they roots, termites, or small, agile creatures. In these conditions, the Neandertals didn't have a hope, lacking as they did the modern human ability to develop the necessary fine-tuned techniques and tools.

But there is no ground for smugness. The traces of these modern humans were likewise lost in the Europe of those days; the DNA specific to all known individuals from the time before the eruption is no longer detectable in our modern gene pool. Mind you, *sapiens* still had a safe haven in the south. The survival of this population, which was spread across almost all latitudes, was – unlike that of the Neandertals – never threatened by the volcano.

The first Ice Age European whose genetic lineage extends all the way to the present was the one who died in Kostenki 39,000 years ago. Markina Gora Man, as he is known, spent at least the end of his life in the western part of present-day Russia and was buried in the layer of volcanic ash, not beneath the volcano itself. This indicates that the man, and hence his ancestors, didn't come to Europe until after the great disaster – perhaps simply because the previous population had been destroyed, or at least decimated, by the consequences of the eruption, thus freeing up the region for newcomers.

At any rate, there is evidence of a pre-existing culture of modern humans reaching from Europe to Asia. The people of Bacho Kiro and Tianyuan Man, who lived 1,000 years later at the other end of the Eurasian continent, evidently used the same tools, as various archaeological finds have shown. This alone points to a degree of kinship. And we know that the Early Upper Palaeolithic culture extended from Israel to northern China. Indeed, a genetic relationship between Bacho Kiro and Tianyuan has now been proven. Somewhere between the two sites there must have been an original population that was dispersed at least over the southern half of Eurasia, but then disappeared under unexplained circumstances. Our ancestors are on the shortlist of suspects.

Outclassed

The speed at which *Homo sapiens* spread across the world is breathtaking: this species overcame the limits imposed by biology on all other animals. Thanks to their newly developed hunting and slaughtering techniques, modern humans were no longer creatures among many others, but the most efficient killers of all time. With blades, lances, spears, and traps, they hunted down everything that tickled their tastebuds and always set their sights on the juiciest steaks, in the form of big game.

The newcomers were no different from Neandertals in this respect – or from Denisovans, for that matter. But they probably went about it with much greater efficiency and subtlety, if you want to call it that. Because, wherever they went, it was never long before the local megafauna was wiped out.

This set our ancestors fundamentally apart from other human species, among which overhunting of the animal world never occurred on such a scale. Even after hundreds of thousands of years of coexistence with the Neandertals, mammoths remained by far the more stable population of the two and, thanks to the existence of hyenas, early humans were not even exempt from demotion within the food chain. Extermination wasn't something the Neandertals were capable of, and so they unintentionally laid the foundation for a sustainable way of life. The trouble was, their way of life didn't stand a chance against the greed of the newcomers.

Soon after the arrival of our ancestors and the disappearance of the Neandertals, the European mammoths went extinct too, to be replaced by incoming Asian mammoths, who likewise never lived to see the end of the Ice Age. Smaller animals of the Eurasian steppes, such as woolly rhinoceroses, giant deer, prehistoric horses, bison, cave bears, cave lions, and cave hyenas became similarly endangered, until they too eventually vanished as a source of food – or competition. Lucrative meat bearers were pursued relentlessly by the newcomers – to the end of the world, if need be (Figure 17).

Not long after setting forth, about 45,000 years ago, or even slightly earlier, according to some datings, humans arrived in Australia, where the same thing happened all over again. Multi-ton marsupials, giant kangaroos and wombats, metre-long lizards and bird species weighing hundreds of pounds were all devoured to excess by the newcomers. Overhunting the available megafauna remained inherent to humans, almost as if they were powerless to resist this self-destructive urge. Even when the ancestors of the Māori became the first humans to colonize New Zealand, some 800 years ago, it took them less than a hundred years to wipe out the very meaty moa birds, up to three metres in height, which lived there in their tens of thousands.

The only parts of the world where megafauna is still present today are Africa and southern Asia – the very same regions where human colonization dates back in some cases to *Homo erectus* and that were already

Figure 17 The northern mammoth steppe © Tom Björklund

Modern humans made repeated incursions into the northern mammoth steppe. Many of these attempts failed, until our ancestors finally settled, 39,000 years ago. Their arrival sealed the fate of the local megafauna, and hence also of the Neandertals.

inhabited by modern humans hundreds of thousands of years ago. What seems to be a blatant contradiction of the theory that the modern human was incapable of hunting big game sustainably is in fact no such thing: precisely because rhinos, giraffes, hippopotami, and elephants 'coevolved' with humans in Africa and Asia, they had time to adapt to the mortal danger posed by these originally quite harmless apes. Fear of humans passed down through the generations in the African megafauna, etched into the animals' DNA. We can see this for ourselves, if we compare the relatively tame wildlife of New Zealand or North America with that of the African steppe, where lions or elephants could still tear us to shreds or trample us to death today if they weren't aware that we might launch a deadly spear or bullet at any moment. As it is, the very sight of a bloodthirsty *Homo sapiens* is enough to make them beat a hasty retreat.

Conquering new hunting grounds was just one way in which humans, in ever-increasing numbers, sought to satisfy their growing hunger for meat. Another was to expand their range of food sources whenever the premium steaks offered by large animals began to decline. From mammoth they switched to deer, and from there to smaller deer, pigs, rabbits, and so on. This required them not only to catch more animals but to go about it with much greater skill; smaller animals are, after all, much faster and nimbler than a ponderous mammoth. To this end, our ancestors developed better weapons, cultivated patience and accuracy, and learned, through sophisticated hunting strategies, to bag even the swiftest prey, whether on land or in water.

Over time, humans added more and more elements to their diet, becoming at once more creative and less fussy. They cracked open the shells of land tortoises over the fire and explored uninviting-looking varieties of seafood; nor did they shy away from insects, spiders, termites, beetles, and maggots as a protein supplement to fungi, roots, and, no doubt, also tree bark. In this way modern humans probably came much closer to the dietary habits of apes than the Neandertals would ever have done. This was the unappetizing price to be paid for *Homo sapiens*' rise and for the maintenance of this species' energy-guzzling brain. And perhaps also for its ability to maintain the taboo on cannibalism even through the toughest times – unlike the Neandertal.

Coevolution

The principle of coevolution also applies to smaller animals. One example is the passenger pigeon, of which a population of up to five billion still existed in North America in the nineteenth century and must at times have darkened the skies of New England. The arrival of Europeans and the invention of gunpowder subsequently put an end to the air superiority of these birds, whose meat was sold for next to nothing. The last one was shot from the skies in 1914.

These pigeons may soon be making a comeback, however. In 2021, George Church – the Harvard geneticist who once announced his intention to resurrect the Neandertal – launched the project 'The Great Passenger Pigeon Comeback'. The declared aim of Church and other renowned researchers is to reconstruct and extract the extinct bird's DNA from old sequences and from the genome of the present-day band-tailed pigeon. Also on the list of this Revive & Restore programme is the woolly mammoth. But even if this project were to succeed and mammoths were one day to roam in zoos across the world, the untold thousands of animal species wiped out at a stroke when our ancestors made their appearance on the global stage will remain irretrievably lost. In the vast majority of cases, we are not even aware of that loss, because they left no fossils behind – let alone the DNA needed to bring them back to life.

The discovery of America

Modern humans conquered habitats previously out of bounds to the Neandertals, the Denisovans, and all other hominins. In 2019 an analysis of DNA extracted from a few milk teeth proved that modern humans had already advanced into bitter-cold eastern Siberia – and indeed above the Arctic Circle – at least 31,000 years ago. Traces of the population labelled Ancient North Siberians were found on the banks of Russia's Yana River, alongside thousands of other signs of human habitation such as stone tools, animal bones, and ivory.

These finds provided a clear indication of these humans' preferred food source: woolly mammoth, woolly rhinoceros, and bison. If this came as no surprise, the DNA analysis certainly did. The Ancient North Siberian lineage branched off soon after the split between Europeans and Asians, yet they share more genetic traits with contemporary Europeans than with Asians. This means that they probably spread eastwards along the Arctic, or else they couldn't have avoided picking up some Asian genes along the way. Consequently the Ancient North Siberians probably inhabited large parts of Eurasia along the

Arctic Circle, clinging on there despite the most adverse conditions imaginable.

Exactly when humans found the way to Alaska across the Bering Strait, which was dry during the Ice Age, in order to colonize North and South America from there, remains a subject of academic debate. There are ambiguous archaeological finds that have been interpreted by some scholars as evidence that America was discovered as early as 30,000 years ago, or occasionally even 130,000 years ago. But all genetic calculations date colonization to 15,000 years ago. The genetic lineages of all Indigenous peoples that live currently between Alaska in the north and Tierra del Fuego in the south suggest that they had common ancestors from this time. Had there been earlier migrations, those first Americans would all have died out.

Most Americans will have probably learnt at school that the Clovis culture of the Palaeoamericans, which 13,000 years ago spanned the entire Midwest of the present-day United States, was the first proof of human settlement in the New World. The evidence for this culture is indeed plentiful, for example in the form of Clovis arrowheads found in bones of mammoths, who were soon annihilated in America as elsewhere. But the 'Clovis first' theory has come increasingly under fire since the 1980s, when the excavation of the archaeological site at Monte Verde revealed signs of human habitation that go back as far as 14,500 years ago.

Recent years have seen scientific consensus grow around the doctrine that the settlement of America must have taken place from north to south, along the Pacific Coast. (According to the classical view, it wasn't until 13,500 years ago that a corridor opened up in the ice sheet that stretched from the Rocky Mountains in the west to present-day Boston, cutting off Alaska from the rest of America.) Even so, it is still hard to imagine humans managing to settle along America's very long west coast, with its many natural barriers, within the few hundred years that must have elapsed between the genetic splitting up of all Native Americans and their arrival in Chile. The latest climate models now raise the possibility of a new alternative; this one is far from garnering scholarly agreement, but should at least be mentioned here as a further option. According to this theory, an ice-free corridor may have formed to the east of the Rocky Mountains as long ago as 15,000 years ago, allowing a passage southwards from Alaska, and hence the colonization of America.

What we do know is that today's Native Americans owe half of their genetic make-up to the Ancestral North Eurasians, who also carried Ancient North Siberian genes. The Ancestral North Eurasians colonized large parts of Eurasia at least 24,000 years ago, probably from eastern Europe to Lake Baikal: consequently their DNA is found not only in Indigenous Americans but also in contemporary Europeans, who received their share during the Bronze Age, when Asian steppe dwellers turned the European gene pool upside down.

The other half of the Native American genome is traceable to East Asian peoples, which must have advanced northwards more than 14,000 years ago and subsequently mixed with the Ancestral North Eurasians. This much can be inferred at least from the DNA that a team from the MPI EVA managed to extract in 2020 from a 14,000-year-old human tooth found in the Baikal region: in terms of genetic composition, this early inhabitant of southern Siberia was indeed very similar to the later Native Americans. This also means that the mixed population of Indigenous Americans did not in fact originate from just across the Bering Strait, as has long been assumed, but probably extended much further south, as far as Lake Baikal.

As for the East Asians' ancestors, they must have kept themselves to themselves for a long period, as the current inhabitants of that part of the world have only a small proportion of Ancestral North Eurasian DNA in comparison with Europeans and Americans. The genetic roots of East Asians can rather be traced to hunter-gatherers from the east of the continent.

To the ends of the earth

Of the Denisovans, who broke away from the Neandertal lineage 500,000 years ago, there is very little genetic evidence, and no skeletons to give us a clue as to their physical appearance. All we have is a few small bones. One is a piece of finger bone the size of a cherry stone, whose DNA allowed the existence of this hitherto unknown prehistoric human species to be proved in the first place. Later on Denisova Cave yielded further bone fragments of Denisovans, and even one from a Neandertal–Denisovan hybrid child.

The only other site where Denisovan fossils have been found is Baishiya Karst Cave, at an altitude of 3,200 metres on the Tibetan Plateau (Figure

18). It was there that a 160,000-year-old lower jawbone was found. A study published in 2019 analysing proteins from the bone together with DNA from the cave sediment indicated that the remains are probably from a Denisovan. This suggests that the Denisovans may have been the first humans to venture into such regions. And they could do so only thanks to a genetic adaptation: a mutation that is still found today in almost all Sherpas in the southern Himalayas and that is credited with enabling them to live permanently at extreme altitudes.

This is a mutation that serves to switch off a function present in almost all mammals: the boosting of red blood cell production in response to increasing altitude and diminishing air pressure. Because falling ambient pressure makes it harder for the lung to activate the exchange of oxygen via the blood cells, the body compensates by generating more red blood cells to move it around. In the short term – when climbing Mount Everest, for example – this mechanism is extremely useful. It's only when you consider the long-term consequences that you realize why this would be a disadvantage for people who actually live at such altitudes. Raising the red blood cell count can also have deadly side effects, including blood clots – which can cause problems such as strokes or heart attacks – and a higher risk of stillbirth.

This is precisely what the genetic mutation in Denisovans served to prevent. Since they could scarcely have survived without sufficient oxygen, another genetic adaptation must have accompanied the mutation in question to make this possible. What that was, we don't know. And so it remains a complete mystery why Sherpas – most of whom also have this mutation – cannot adapt to altitude like all other humans, yet have no problem at all escorting the rest of the world to the earth's highest summit when required. After their ancestors inherited this mutation from the Denisovans, it remained dormant for a long time, only appearing in a very small fraction of the population; it was not until 35,000 years ago, according to analyses of up to 4,000-year-old genomes from this region, that it began to spread among the inhabitants of Tibet, which probably coincided with their advance into loftier habitats.

In other words, the Denisovans might have been far more capable than the Neandertals of adapting to a wide range of environmental conditions. The two direct genetic proofs of the Denisovans' existence are thought to have originated from the northern geographic range of these early humans; thus this area may have extended across the entire mountain region from Mongolia through China to the Himalayas. Our ancestors seem to have had little contact with this population though: contemporary Chinese and other peoples of Southeast Asia carry only about two parts per thousand Denisovan DNA. It's a different story in the neighbouring continent of Oceania. There the DNA of the Indigenous people testifies, even if indirectly, to the Denisovans' ability to settle not just in mountain environments but in coastal ones too.

Last stop Tasmania

Today Denisovan DNA is still found in small quantities in Chinese, Korean, Japanese, and also Native American populations, but it is far more prevalent in New Guinea and Australia. There the Indigenous people carry not just the roughly 2 per cent Neandertal DNA common

Figure 18 The Tibetan Baishiya Karst Cave © Dongju Zhang / CC BY-SA 4.0

The Tibetan Baishiya Karst Cave lies at an altitude of almost 3,300 metres and is thought to have been inhabited by Denisovans as early as 160,000 or more years ago. Denisovans appear to have been particularly well adapted to high altitudes – a genetic characteristic that modern-day Sherpas still benefit from.

to all non-Africans, but also around 5 per cent Denisovan DNA. The lineages that led to Oceania and East Asia were separate, however: according to genetic calculations, the Denisovans must have split into at least two populations some 200,000 years ago. This gave rise to northern and southern Denisovans, who differed in their DNA but still belonged to the same group of *Homo*. The northern branch lived in the Altai mountains and Tibet, while the southerners appear to have mixed with modern humans en route to New Guinea.

Although there is no ancient DNA from southern Denisovans to prove this, we do have those very clear traces in the genomes of Indigenous New Guineans and Australians. This begs the question of whether modern humans first mated with the Denisovans in New Guinea or had picked up their genes beforehand. The answer is anything but irrelevant: it would give us a clue as to whether the Denisovans were already a step ahead of modern humans in the conquest of Oceania – or failed at the last hurdle, which we went on to overcome: water.

During the Ice Age, Australia and the island of New Guinea to its north formed a continuous landmass, known as Sahul. On the other side of the water lay the region of Sundaland, joined to the Asian mainland. Between the two landmasses was a group of islands named Wallacea, one of the most impenetrable biological barriers in the world. To this day, the flora and fauna on either side of Wallacea remain fundamentally different as a result of strong currents that prevented any kind of species exchange here for millions of years. To reach Sahul, this barrier had to be overcome. But the Denisovans may well have managed to navigate their way across. The next barrier was the greatest of all: the open sea at the end of the Indonesian Archipelago. For, no matter from where one decided to cross this vast body of water, Sahul remained invisible, being about as far away as Helgoland is from the North Sea coast. Anyone attempting such a gamble could only hope to make it to the other side alive.

Our ancestors pulled it off: that much we know. Whether they took Denisovan DNA with them on their epic voyage or picked it up on Sahul itself must, for the time being, remain a matter for speculation. Perhaps the Denisovans were bold enough to attempt this major venture before our ancestors – who knows? For, alas, their faint traces were lost in the DNA of Oceania, being detectable only in a tooth from the Altai

Mountains and in a lower jawbone from Tibet. We know neither how long the Denisovans lived, nor how far they got, nor how they disappeared. All we know is that they disappeared, just like the Neandertals.

For the kamikaze band that finally conquered Sahul about 45,000 years ago, that remained a lonely expedition. After the first modern human migration to Australia, there were no further attempts, at least as far as we can tell from the DNA of Aborigines today. Over the millennia, the Aborigines managed to develop an impressive hunter-gatherer culture with its own brand of mysticism and religion, carving out a home for themselves on an inhospitable continent. With the end of the Ice Age, however, an insurmountable barrier formed between their homeland and New Guinea, while Tasmania in the south also became isolated from Australia.

From then on, the Tasmanians had no more contact with the outside world, and archaeological remains show that their culture and technical innovations were lost over time. When Europeans rediscovered the island 11,000 years later, the ancient relatives they found there had only the most primitive stone tools. The fate of the Tasmanians was sealed: massacres, imported infectious diseases, expulsions, and, ultimately, systematic slaughter saw to it that the Indigenous Tasmanians were almost exterminated.

For tens of thousands of years, southern Tasmania was the remotest corner of the earth that our boldest forebears had ever reached. In 1905, in the small town of Cygnet in the southeast part of the island, the final traces of this vanguard of humanity dissolved in the DNA pool of European migrants: that was the year that saw the death of Fanny Cochrane Smith, the last descendant of two unadmixed Indigenous Tasmanians.

6

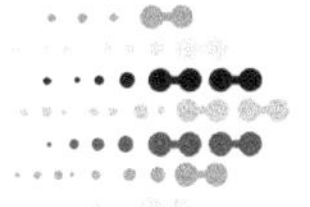

Enchanted Forests

About 40,000 years ago things start to get hairy in Southeast Asia. Here we encounter rainforest, hobbits, and terrifying beasts. Down under, in Australia, it gets even worse. But we press on regardless. Now and then we are forced to swim, without knowing where to. One of us will make it through eventually. Meanwhile they are breeding dogs in the north.

A fearsome place

Humans' understanding of their own origin has less to do with their actual past than with the traces they leave behind. This is something we should constantly bear in mind when reconstructing the colonization of the planet by our ancestors. Archaeology is reliant on remains of human settlements – on prehistoric firepits, tools, and skeletons. Even more so is archaeogenetics, which draws on bones, teeth, and hair in which DNA has been preserved for aeons. The chances of useable – or even discoverable – relics of human life surviving for posterity are favourable when they remain buried in the ground, cool, dry, and shielded as far as possible from UV rays – not in places where damp heat and bacteria cause them to decay. Also, archaeology exists where there are archaeologists. This branch of research was born in eighteenth-century Europe, and this is where it came of age. This, too is a reason why there is hardly a cave on the old continent that hasn't been scoured for Stone Age fossils by inquisitive researchers and amateur archaeologists. The tendency of archaeogeneticists to focus on finds from cooler, temperate, and dry latitudes is reinforced by the climatic conditions there: cold dry air acts like cottonwool on the fragile DNA structures in human remains.

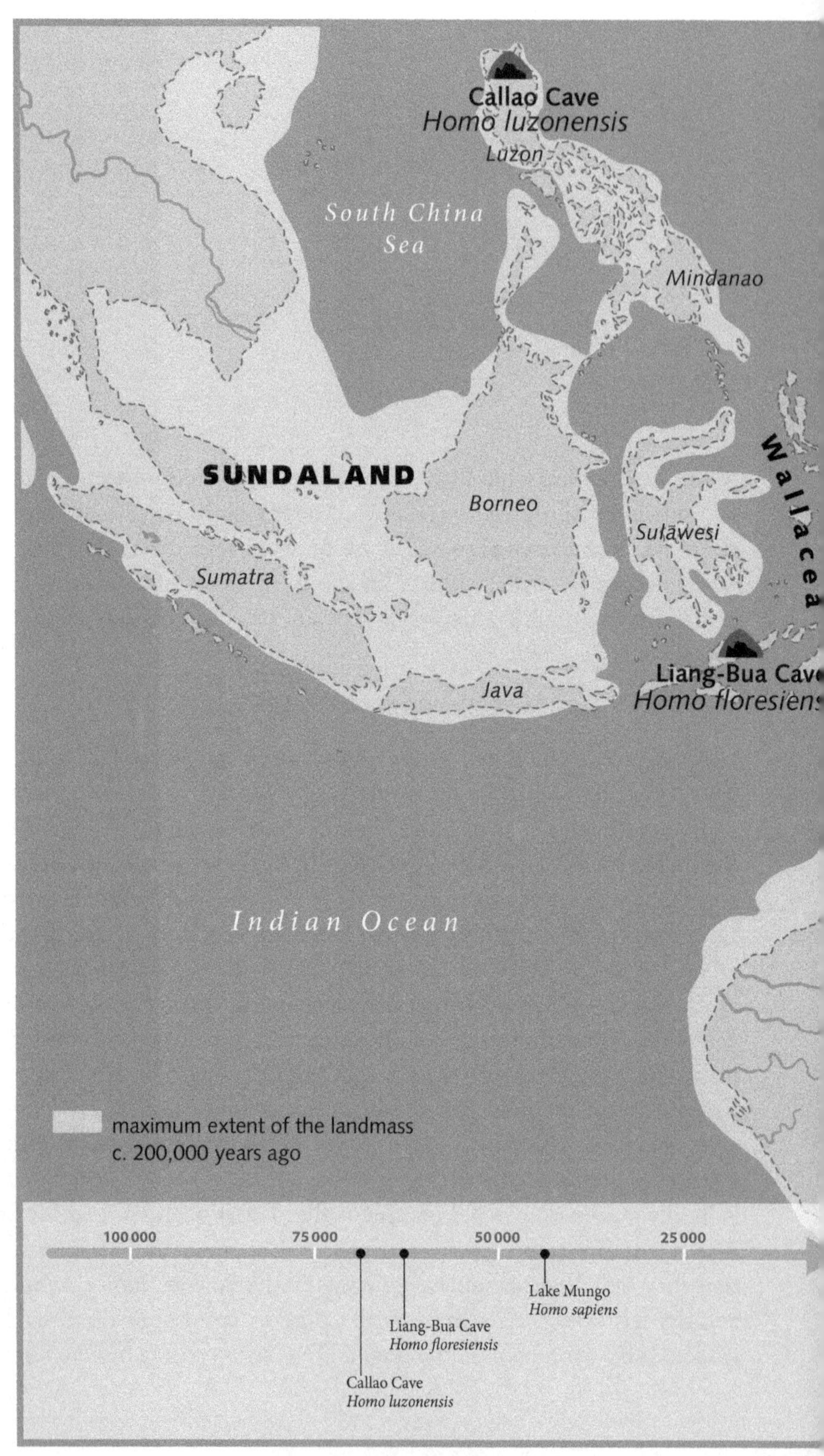

Figure 19 Enchanted forests.

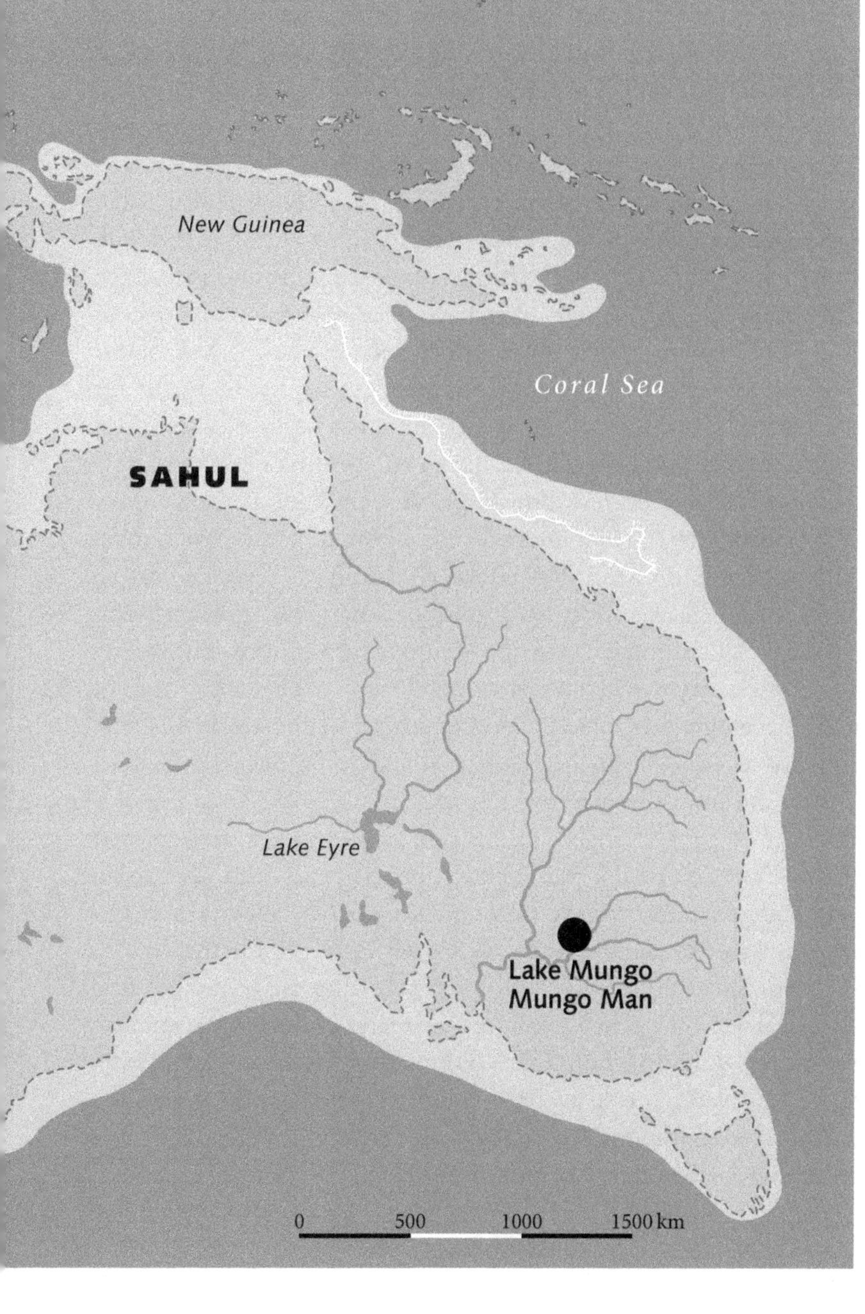
Pacific Ocean
New Guinea
Coral Sea
SAHUL
Lake Eyre
Lake Mungo
Mungo Man
0
500
1000
1500 km

The northern part of Eurasia thus offers better natural conditions for archaeologists and archaeogeneticists – a fact that is compounded in Europe by a historical overemphasis on local traces of human migration. From the African jungle, where the common ancestors of the great apes lived, we have, by contrast, precious little: bones and other remnants of deceased individuals couldn't survive more than a few days, let alone millennia. That said, a peculiarity of the eastern flank of Africa has resulted in an accumulation of finds in an otherwise largely dry region: the Great Rift Valley was not just a corridor used by people in the east of the continent as a main route north- and southwards – it also offered ideal conditions for excavating their remains even millions of years later. The constant movement of the African and Arabian plates as they rub against each other and drift apart again has exposed more and more rock strata that archaeologists would never have been able to access without the aid of plate tectonics. It is quite a different story in the Sahara, for example, whose settlement history during its green phase remains for the most part a closed book. There is virtually nothing to excavate there – just a lot of sand. Anyone who has lost a ring on the beach and only noticed the next day will know how hopeless such a task is.

All the Neandertal, *Homo erectus*, and, ultimately, modern human finds in northern and especially western Eurasia have depended on an excellent archaeological infrastructure and ideal natural conditions. This is not to diminish the importance of the story as we know it, but merely to put it in perspective. We have mentioned some of the deficiencies of African archaeology, but there is another region of the world that has been similarly overlooked in archaeological and archaeogenetic research in the past. And unjustly so, as we now know.

It is becoming increasingly clear that the subtropical and tropical regions of East Asia, and particularly the belt between the Asian mainland and Australia, deserve closer scrutiny. Archaeological remains in this region are still sparse, and DNA even more so, for the reasons already given. But the little we have paints an impressive picture of a quasi-mystical world in which our ancestors made a home for themselves in the rainforest – the antithesis of everything that had made their predecessors great. It was a world whose inhabitants lived on giant rats and were surrounded by dragon lizards, who in turn fed on elephants. The paradisical archipelago that stretches today from the Asian mainland to

Australia was then a fearsome place – and an epicentre of early human diversity, far away from the African cradle.

Endless rain

For one thing, the jungle is no place for hunter-gatherers. It was not without reason that the ancestors of *Homo erectus* had to flee the rainforest before they could develop the elevated head position they needed in the African savannah, which in the end turned them into avid carnivores with huge brains. Even for skilled hunters, the primaeval forest is a dead loss: it offers them no overview, too little freedom of movement, and, above all, far from ideal conditions for killing an animal at long range with spear, lance, or bow and arrow. The potential prey, on the other hand, has a wealth of hiding places, so that endurance hunting is simply not an option.

It is not impossible for humans to survive in the jungle, and there are still some scattered Indigenous peoples who are proof of this. However, these few remaining populations are often part-time farmers who habitually opt for slash-and-burn clearance as a means of carving out some arable land from the rainforest. Our ancestors, who went on to conquer the Southeast Asian jungle, knew nothing of farming and probably nothing of slash-and-burn techniques either: they simply adjusted to the dense undergrowth into which they had ventured from the flat steppes of East Asia – an ability that *Homo erectus* apparently didn't possess. Only *Homo sapiens* would become Tarzan material.[1]

We have long known of the existence of very early hominins in Southeast Asia. One of the few archaeological finds from this region is also one of the first in the world. In 1891, on the island of Java – then part of the Dutch colonial empire and now part of Indonesia – the doctor Eugène Dubois excavated the remains of what became known as Java Man, just thirty-five years after the discovery in Germany's Neander Valley of the first ancient human to be anatomically related to modern man.

Although the condition of the bones renders dating highly unreliable, Java Man appears to have lived around 1.5 million years ago and more than 5,000 kilometres south of another very famous East Asian *Homo erectus*, namely Peking Man, whose age is estimated at around 600,000

Figure 20 *Homo erectus* in Sundaland © Peter Schouten

In the days of *Homo erectus*, large parts of Sundaland offered similar conditions to Africa's – a perfect environment for these early humans, until the jungle began to spread across the region, thereby paving the way for our ancestors.

years. This prehistoric subspecies spread as far as the Pacific coast of Asia, advancing ever further across the steppes and grasslands. There it enjoyed a rich food supply, akin to that of its African home. This means that he probably didn't have to make extraordinary adaptations in order to make it to the Far East.

In the case of Java Man, this was until recently a moot point. Nowadays Southeast Asia is one of the most densely populated areas in the world. But, before this became possible, humans had to push back the rainforest, which still covers large parts of this archipelago to this day. If similar conditions already existed there at the time of Java Man's expansion, that would indicate an exceptional adaptability on the part of this *Homo erectus*. Thanks to recent discoveries, we now know better: like Peking Man, Java Man appears to have relied on finding familiar terrain. Unlike our jungle-conquering ancestors, his descendants were unable to cope with the dramatically changing environment of 100,000 or so years ago.

In 2020, a collaboration between the MPI in Jena and Griffith University in Queensland, Australia proved, on the basis of a biochemical analysis of soil samples from Southeast Asia, that at the time of *Homo erectus* this region looked very different from the place we know today (Figure 20): vast grasslands covered Sundaland, which extended as far as the Philippines, and all kinds of megafauna roamed over the land – such as the stegodon (an extinct relative of the elephant), various types of rhinoceros, and the notorious hyena. For the prehistoric humans newly arrived from Africa, this was like an invitation. In just a few millennia they must have spread along the coast all the way from the Arabian Peninsula to Southeast Asia, via the Indian subcontinent, finding similar conditions everywhere.

But at some point that cosy life in the Far East came to an end: the rain arrived and never stopped. The savanna turned into an ever denser forest landscape in which *Homo erectus* seems to have lost its way. According to biochemical analyses, the grassland was increasingly displaced by woodland, until the rainforest took over, some 100,000 years ago. With the spread of this tropical landscape came the extinction of most of the old familiar animal species. *Homo erectus* ran out of meat, and this species' attempts to cultivate an alternative lifestyle in the jungle evidently failed. Not so modern humans, who set out from the African savannah more than 100,000 years ago and fought their way through the Southeast Asian jungle, ending up at Lake Mungo in southern Australia, 40,000 years ago.

Tough little hobbits

Thanks to the hot and humid climate, no bones from our prehistoric ancestors containing sequenceable DNA have been found on the tropical islands to date. But there have been in the recent past some spectacular finds that relate neither to *Homo erectus* nor to *Homo sapiens*. The people behind them inhabited these islands until around 60,000 years ago. One particular fossil – that of *Homo floresiensis* – was discovered on the Indonesian island of Flores in 2003, the very year in which the third part of *Lord of the Rings* hit the cinemas; and it is the hobbits, barely 120-centimetre tall creatures central to this trilogy, that gave the prehistoric humans their nickname, since they were just as small. While *Homo floresiensis*' feet were not big and hairy like those of the hobbits in the film, they were unusual in another way: they resembled those of great apes, suggesting that these prehistoric creatures may have been able to climb trees – another obvious advantage in the tropical rainforest.

Homo luzonensis, meanwhile, lived almost 3,000 kilometres further north, on the island of Luzon in the northern Philippines: like the hobbits, this species, too, was rather small in stature. A dozen bones and teeth from this *Homo luzonensis* were found in a huge cave in 2013, but no DNA could be distilled from them. It is still not clear whether Luzon Man was just another hobbit, or in fact a separate hominin species. Nor can we, in the absence of any sequenceable DNA, rule out the possibility that both were a variety of southern Denisovans. The latest DNA analyses suggest that this is quite possible, at least in the case of *Homo luzonensis*. According to a study published in August 2021, the Indigenous inhabitants of the island of Luzon – the Aeta people – carry 30 to 40 per cent more Denisovan DNA than Australians, for example (Figure 21).

Both the hobbits and *Homo luzonensis* appear to have survived the mega-eruption of the Mount Toba volcano. This is not surprising, given that both the northern Philippines and the island of Flores are about 2,800 kilometres away from Mount Toba as the crow flies, and hence further from the scene of the disaster than the belt between East India and Vietnam. In the circumstances, we can imagine that in the apocalypse after the volcano the modern humans who had made their home in the Asian jungle were more severely deprived of basic livelihoods than the smaller humans who lived amid the luxuriant vegetation of the rainforest.

Smashed skulls

After the discovery of the hobbit in 2003, a long debate ensued as to whether this was simply a modern human with a distinctive pathological mutation – in other words, one of our ancestors with a genetic defect that produced a conspicuously small skull. Speculation centred on the possibility of microcephaly, a condition that accounts for a conspicuously

Figure 21 Luzon Man © Michael Weis / Alamy Stock Foto

Luzon Man lived in vast caves, one of which now houses a chapel.

small head and is often associated with limited mental capacity. Since all other hobbit finds point to the same body size, this is rather unlikely: it would be a huge coincidence if the few finds all came from modern humans who suffered from a rare malformation.

We know little about the physical appearance of hobbits apart from their small stature and approximate skull size. This is partly due to a rather tragic saga of discovery and reconstruction. When, after some back and forth, the skull and a fragment of pelvic bone from the type specimen had to be transported between Indonesia's two main rival archaeological institutes, the brittle objects shattered into hundreds of pieces. After that, their shape could no longer be reliably reconstructed: in the case of the hip, there was no hope of restoring any part of it to its original state, while the skull had to be stuck together in makeshift fashion. This certainly impacts any analysis of whether the hobbit's anatomy is suggestive of a modern human, or indeed a Denisovan. The hope remains, however, that bones that contain useable DNA – and hence are suitable for ancestry analysis – will be discovered in the coming years. Other bones have also been found on Flores since the skull and pelvic bone. These, too, suggest that *Homo floresiensis* was specially adapted to climb trees – a finding that rather contradicts the notion of a family relationship with modern humans or Denisovans.

The spaces once inhabited by hobbits and by *Homo luzonensis* are still an awe-inspiring sight. The Callao cave system where the latter was found consists of several gigantic chambers, each large enough to house a church – and the locals have indeed turned one of them into a chapel. Apart from tourists and churchgoers, the caves are now colonized by all kinds of creatures, notably hundreds of thousands of bats. The diminutive *Homo luzonensis* must have seemed even more lost in these vast spaces than our own kind, yet managed to survive there for several hundreds of thousands of years. And *Homo floresiensis*, who lived at the other end of the Southeast Asian rainforest, inhabited an environment similarly reminiscent, in places, of Tolkien's fantasy world.

The species' homeland of Flores had been occupied for at least a million years – or at least that's how far back the archaeological trail leads. *Homo erectus*, Denisovans, *Homo sapiens* – all might have got this far. Whoever it was that made it onto the island reached Wallacea, that great barrier that proved insurmountable for all other animal and plant species. Moreover, Flores was not part of the southeastern landmass of Sahul, but an island – then as now: perhaps this is why its wildlife was like nothing humans were likely to have encountered anywhere else on the planet. The thought of trying to get by here as a pint-sized representative of the genus *Homso* is positively hair-raising.

To this day, the island is still inhabited by rats the size of beavers. Countless bones from these almost metre-long creatures have been found in the caves occupied by hobbits, which indicates that they must have fed chiefly on rats almost as big as themselves. Then again, neither rats nor humans were much bigger than the dwarf elephants of Flores: this smallest known and now extinct species of Stegodon reached a shoulder height of approximately 1.20 metres and weighed around 300 kilos. It was hunted by humans, though far more commonly by giant lizards.

Known as Komodo dragons – after the island of Komodo, to the west of Flores – these lizards live nowadays on a few islands in the region, to the delight of tourists but to the dismay of the locals, who are occasionally attacked and sometimes even killed by the animals. These reptiles, now threatened with extinction, are up to three metres long and can move very fast. They are aptly named, too, as they resemble flightless dragons and are equipped with a venom gland that they use to subdue their prey before shaking it to death and tearing it apart.

And that's just the quick version: there is yet another, far more agonizing way of being killed by a Komodo dragon, and that is to be bitten by it. This causes a mass of lethal bacteria to be transferred from the dragon's saliva into the victim's bloodstream. It can then take days for the cocktail of venom and bacteria to do its deadly work. The dragon, which can smell its weakening prey from several kilometres away with its deeply forked tongue, pursues it until it finally collapses. Among the hobbits, this giant lizard topped the list of most feared predators. Luckily for them, however, it preferred to feast on the far meatier mini-elephants.

An isle of dwarfs and giants

This island landscape, which appears so grotesque to us today with the likes of giant storks three metres in height, turned the usual proportions between species on their head. There are genetic explanations for this, and all have to do with an insular geography to which humans and animals evolved to adapt. The small stature of hobbits was probably an accident that couldn't be subsequently remedied – the island could just as easily have been populated by outsized humans. For whenever it was that the hobbits' ancestors came to Flores, they evidently brought with them genetic characteristics that, thanks to the evolutionary bottleneck effect, produced a whole population of subsized humans. It may have started with a single original hobbit who was a head or two shorter than his peers but whose children, for some reason, had a better chance of survival and went on to produce another generation of subsized people. Had there been no further migrations to the island, this initial wave might have paved the way for a hobbit reign that could have lasted tens of thousands of years (Figure 22).

The animals on Flores, meanwhile, evolved the way they did because they could – or had to. Unlike the rest of the world, where rats were and still are considerably smaller than the specimens on Flores, this island lacked sufficient rodent predators. Since most mammals keep growing

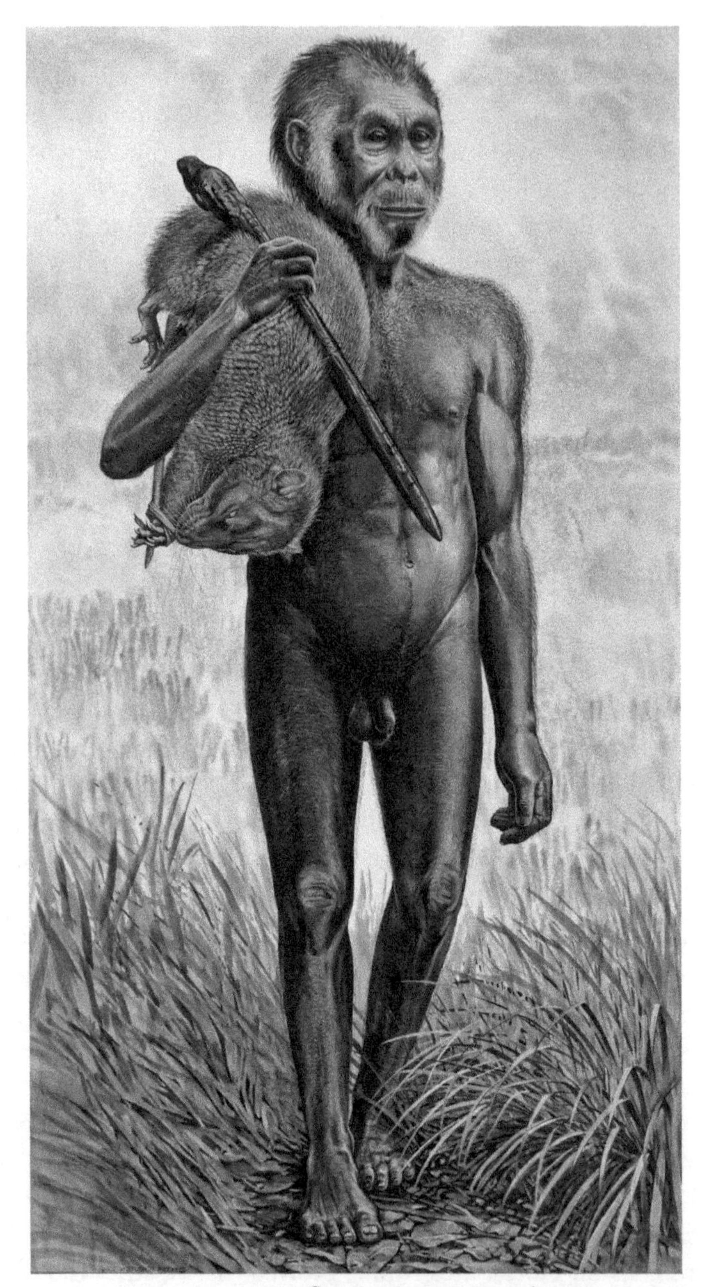

Figure 22 *Homo floresiensis* © Peter Schouten

The diminutive *Homo floresiensis* lived on the island of Flores among very large animals. One of these was a giant rat that measured 1 metre from nose to tail.

until they reach sexual maturity, small-bodied individuals are favoured when there are many dangerous predators around, as they can breed a large litter quickly and thus secure the survival of their species before they become prey. Consequently rats, which are hunted by humans and animals alike, are relatively small the world over and reproduce at a fast rate.

On Flores, by contrast, those rats that reached sexual maturity later and thus were larger and stronger had a competitive advantage in terms of reproduction. This appears to have outweighed the evolutionary disadvantage posed by selection through predators – in other words hobbits and giant lizards. As a result, the small Asian rat was able to live relatively unmolested and to get bigger and bigger on Flores than in the rest of the world. Whether later on the hunting practised by hobbits led to a further evolutionary drift towards smaller specimens remains open to question; thus the giant rats might have been even larger before the arrival of these or other humans.

The limits of growth

On an island, the trend towards maximization is not endless, but depends on many other evolutionary parameters, notably food supply. The food sources of land mammals will always be restricted, thus imposing a natural limit to growth. This applies to elephants even more than to rats, of course: on an island like Flores, the original version of their species would struggle to obtain the hundred kilos of vegetation they need every day – at that rate, even the lushest rainforest would eventually be stripped bare. In these circumstances, individuals who mature and produce offspring sooner have a clear advantage: only then can a sufficient food supply be guaranteed for the whole voracious horde. Hence the gradual downsizing of elephants on Flores.

In Asia and Africa, on the other hand, such a trend would have been fatal for the local elephants. For one thing, the grazing was unlimited there. And, secondly, any species of elephant reduced to the size of a wildcat would soon find itself in the latter's sights:

after all, the pachyderm's huge bulk is its main protection in the wild, keeping it safe from attack by tigers and lions. Although the dwarfism of Flores' elephants made them a favourite on the Komodo dragons' menu, here too the evolutionary rule of thumb applied whereby whichever selective advantage entails the fewest disadvantages is always the one that wins out.

Eurasia a mere sideshow

It was a combination of genetic accidents, limitations, and possibilities that gave Flores – and perhaps also Luzon – a quite unique insular character. The fact that the hobbits, whose main prey must have been half their height if not more, managed to survive for a very long time and apparently remained at the top of the food chain despite natural conditions that were less than favourable to them, is further proof of the evolutionary superiority of the genus *Homo*. Like the Neandertals, the Denisovans, and *Homo erectus*, hobbits too were able to compensate for their physical weakness, which was striking when you compare them with other predators; and they did so through intelligence, which they used to develop weapons and hunting strategies and techniques. But, as we now know, this wasn't enough in the end: the hobbits vanished from the face of the earth some 50,000 years ago, as did *Homo luzonensis*. Whether our ancestors had anything to do with the disappearance of these early South Sea people, we don't know – all we know is that they, too, arrived in this region just over 45,000 years ago. And they finally succeeded where all their predecessors had failed, settling permanently in the rainforest.

That said, the Denisovans may have been on a similar path before them. They appear to have inhabited these parts up until 25,000 years ago, or so we gather from a Denisovan lineage among the genes of contemporary New Guineans – a lineage that exists in addition to those found in all East Asians and that must have been a later admixture. Analyses conducted in the past few years show that our ancestors probably shared this vast and fertile territory with the Denisovans for thousands of years. The magical island realm of Southeast Asia, which today attracts tourists from around the globe, may have one of the most formidable natural barriers in the world running through it, but that never deterred the genus *Homo*. For tens of thousands of years, Southeast Asia was home to a variety of human types that were probably all but non-existent outside Africa. Wallacea – the region that begins east of Bali and Borneo and stops at New Guinea – was a central hub of human expansion. For a long time, northern Eurasia was just a sideshow.

We don't know how the hobbits came to Flores. But it must have been by sea, since even in the depths of the Ice Age there was not enough water frozen at the poles and in the Eurasian glaciers to turn the whole Southeast Asian island chain into a land chain. Since the hobbits, like other early humans, were presumably unable to swim long distances, they must have come over on tree trunks, or perhaps even on primitive rafts. This would have been quite a feat, especially beyond Bali: depending on the water level at the time of colonization, the distance to the other side would have been 20 kilometres or more.

A nightmare voyage indeed. Nowadays at least, the currents in these waters are deadly – a fact exploited by many a diving adventure company in Bali whose clients are cast into the drift, to be carried at breakneck speed past the coral reef and retrieved several kilometres downstream (a fair number apparently don't make it). No early human would have embarked on such a suicide mission just for the adrenaline rush. Rather the hobbits' ancestors must have run out of space in Sundaland or on the offshore islands; or perhaps they came under pressure from incomers such as the Denisovans. Another possibility, as already discussed, is that the hobbits – for whom we have no genetic data – were themselves a branch of the Denisovans: a dwarf version, so to speak, of the same prehistoric human species.

Deadly Australia

Just thinking about how Stone Age humans managed to migrate to Flores boggles the mind. How much more so, then, in the case of Sahul – the landmass that became New Guinea and Australia at the end of the Ice Age.

At what point our direct ancestors took to the water before ending up in Sahul is not clear. One possibility is the southern route, the Ring of

Fire from Sumatra to New Guinea via Bali, Lombok, Flores, and Timor. The other is the northern route from Borneo (which was then part of the Sundaland landmass) via Sulawesi. Current research leans towards the northern route, as there was less water to cross there and more closely spaced islands that could serve as stepping stones along the way. But even this must have been anything but a pleasant South Sea excursion: many pioneers in search of new hunting and gathering grounds will have ended their days on the ocean floor or as fish fodder.

Those who made it began by conquering the northern part of Sahul, now New Guinea. The descendants of these first migrants still live on the island to this day. As members of more than 300 different Indigenous groups who speak over 800 languages, many still practise hunting and gathering, alongside a form of slash-and-burn agriculture. Cannibalism, too, persisted into the last century among some Indigenous people: according to contemporary sources, this was a cult ritual designed primarily to harness the power of slaughtered enemies by eating their brains and other body parts.

The further south our ancestors advanced, the more hostile the conditions became. The humid tropical climate of the Equator gave way to a deadly aridity that still persists today in large parts of Australia. Our ancestors avoided the grassland and desert of central Australia, as is still detectable today from the genes of their descendants, the Aborigines. Their mitochondrial DNA – the part of the gene inherited through the female line – can be traced back to a common female ancestor who arrived in northern Australia around 45,000 years ago. The subsequent divisions of the mitochondrial DNA found in contemporary Aborigines chart the millennial history of human settlement on Lake Mungo in the southeast of the continent. The migrants spread out in a ring, starting with the tropical and temperate parts of eastern and western Australia. Although some also ventured into the deserts and steppes of central Australia, this remained peripheral territory rather than a place for them to settle permanently. It wasn't until modern European migrants came along that the Aborigines were driven into the Outback – if their lives were spared at all.

The flora and fauna of Australia continue to fascinate visitors today. When the first humans arrived there, most of what they found was highly deadly. On a continent that sometimes sees no rain for years, only to be deluged by sudden monsoons, evolution favours species with plenty of staying power, a reputation for toughness, and, above all, the ability to avoid ending up as someone else's dinner. For this reason, Australia has more venomous than non-venomous snakes, not to mention deadly spiders and scorpions, or the only mammal in the world with a venomous spur: the platypus. And things were even worse for humans at sea, where the numerous sharks were (and still are) the least of one's worries: deadly dangers lurk in this space in the form of octopuses, stonefish, sea wasps, box jellyfish, and cone snails. The first humans to set foot in this hostile place may well have wished themselves back among Southeast Asia's fauna, which, if not necessarily less dangerous, was certainly less devious.

Leaving this aside, Australia offered rich hunting grounds. There humans encountered the largest marsupial in the world, the herbivorous Diprotodon. This creature grew to a height of around two metres and resembled a giant wombat. The marsupial lion was larger than African and Eurasian lions; ditto the Australian thunderbird, which could weigh half a ton and more. As for the Megalania, a species of giant lizard, it weighed more than a ton and probably preyed on marsupials and smaller reptiles. These huge animals met the same fate as most others around the world: when humans came, they went. The drama that began in Eurasia continued seamlessly Down Under.

Australia yielded a further example of an anthropological constant that has been demonstrated in the meantime in the Southeast Asian islands as well: whenever the first human tools appeared, the megafauna – except in the case of Africa and Southeast Asia – duly disappeared. The arrival of humans transformed the region far more radically than the change in climate 100,000 years ago, which turned the grasslands to rainforest.

Hunting with wolves

Since they first learned to think, humans have, it seems, divided the wildlife around them into three main categories: meat bearers – to be eaten, and hence killed, in as large numbers as possible; dangerous animals – to be kept at bay, or simply killed; and non-dangerous non-edible animals – to be tolerated or, if they become a nuisance, killed. Within just a few thousand years, our ancestors had wiped out all the megafauna from Eurasia to Australia. Only then did they come to discover an entirely new way of interacting with animals: coexistence and use. It was still a long time before they began to domesticate wild animals, however. And perhaps it's just as well that the first experiments didn't take place in Southeast Asia, with the breeding of giant milk and meat-bearing rats, but further north, with the Asian and European forerunners of sheep, goats, cows, and pigs. But before that there was one animal that Stone Age humans of Eurasia made not just a servant but a loyal companion: the dog.

Today there are half a billion domestic dogs in the world, and they consume vast quantities of food. A few years ago, an American researcher calculated that the country's dogs alone, were they to found their own state, would have the fifth largest meat consumption of all nations of the world. By keeping dogs – and cats, of course – humans are increasing their own resource footprint even further, for the sake of animals that serve no particular purpose apart from their owners' gratification.

With hunters it's a different story: for them a dog is usually not just a member of the family but an essential partner. The quadrupeds help them to sniff out, flush out, chase, and sometimes also kill wild animals. It was this unique ability that brought wolves and humans together. And it was the fact of being more useful to us alive than dead that allowed the descendants of the wolf to lead what is probably the most privileged existence of virtually all animals – if you disregard problems such as overbreeding, dog fashion shows, lack of outdoor exercise in urban homes, and the eternal competition with the house cat for human attention.

Exactly where the first humans hit on the idea of taming wolves and domesticating them over several generations is unknown, but it was somewhere in Eurasia, most probably 20,000 to 15,000 years ago. There are also archaeological finds dating back much further that some scholars hold to be early evidence of attempts at domestication. One example is a 33,000-year-old set of dog-like bones found in a Siberian cave that was also inhabited by humans – not that this is necessarily proof of coexistence. On sequencing this animal's DNA, scientists found that it clearly belonged to the wolf family, with no genetic relationship with the dogs of today. Another, very similar example was found in East Belgium. This one was about 32,000 years old and was discovered near a cave that was inhabited by humans at the time. It, too, had no genes in common with modern dogs.

Both these finds raise at least the possibility that humans across Eurasia had already begun to cultivate the wolf's affections at that time. But it was not until the discovery of the skeleton of a dog buried some 14,000 years ago, together with his master and mistress, in what is now Oberkassel, near Bonn, that the first genetic traces of modern dogs were found (see Figure 23). What we can say is that, according to genetic calculations based on modern canine genomes, the last common ancestor of all our four-legged friends lived about 20,000 years ago.

The genetic diversity of dogs is extremely narrow. They are thought to originate from a population in the four-figure range, but it is equally possible that the hundred-plus modern breeds go back to just a few hundred primitive dogs. This has resulted in an incredibly low level of genetic variation in modern dogs. The DNA of two random German Shepherds, for example, will be as similar as that of first-degree relatives in the human world – a fact responsible for the prevalence of congenital hip dysplasia in this breed.

That said, present-day wolves are also descended from a comparatively small initial population of roughly 10,000 individuals: their last common ancestors lived around 60,000 years ago. This means that there must have been a dramatic explosion of the wolf population around the time when modern humans came to Europe. These animals moved into the ecological niche vacated by our ancestors: the caves previously occupied by hyenas. Humans probably accepted their new cohabitants because, unlike hyenas, the wolves left them largely in peace, contenting themselves with stripped bones and serving to keep other animals at bay. Among these were brown bears, for example, which also saw a huge population expansion after the arrival of humans, because they were able to penetrate the niche of the cave bear, which died out around 26,000 years ago, along with the rest of Eurasia's hunted megafauna.

The romance between wolf and human gradually blossomed, the wolf possibly making the first moves. This animal is very similar to the human in one respect: it prefers long-distance running and covers the same sort of distances. When our hunting ancestors chased their prey to exhaustion, wolves were able to keep up with them. When an animal was dismembered on the spot, these hungry companions may have

Figure 23 The prehistoric dog © Landesamt für Denkmalpflege und Archäologie Sachsen-Anhalt / Karol Schauer

One of the oldest traces of a prehistoric dog is 14,000 years old and was discovered in Oberkassel, near Bonn. Eurasian hunter-gatherers are thought to have begun taming the wolf much earlier, in order to use it as a hunting assistant.

shared the remains among themselves. Human hunters were probably followed by hyenas as well as by wolves, but those were more likely to be rivals than passive or indeed useful partners. There were, then, many conceivable reasons for killing the hyenas and sparing the wolves.

Tameness is also a mutation

The domestication of the wolf must have begun each time with child predation. Humans probably killed the mother as soon as her young were weaned, then raised the offspring themselves. There must have been several such attempts in different places over time, and no doubt many will have ended in tragedy – sometimes for the breeder, if they were bitten by a wolf, but far more often for the more aggressive animals, who paid with their lives for any attack on their master or mistress. Over the generations, the tamer, more trusting animals will have prevailed and been crossed with animals of a similar temperament – perhaps from other hunter-gatherer groups. That way the tameness of the modern dog could be cultivated in a relatively short time, over decades rather than centuries. This characteristic is genetically determined, as we know thanks to a long-term Soviet experiment aimed at breeding domestic foxes.

Dmitri Belyayev, a Russian geneticist, embarked on this large-scale project in 1959. His purpose was to understand how the wolf gave rise to the dog. His working hypothesis that the character traits of dogs loyal to humans were the result of a deliberate selection process subsequently proved to be spot on. By way of demonstration, Belyayev simulated a speeded-up version of evolution using a group of silver foxes purchased from a Canadian fur farm. He mated with each other the foxes that showed less fear of humans and less inclination to bite, and the process was then repeated in each successive generation.

This way Belyayev managed within ten to twenty generations to breed veritable lap foxes, which wagged their tails and licked the scientists' hands when they approached. Also, the tamer the foxes became, the more they developed physical traits generally perceived by humans as cute, such as floppy ears, curled tails, and shorter muzzles. These are not objectively 'cute', however: they are simply traits that appear to be fostered by domestication and signal to the human eye – conditioned as it is by evolution – that the animal in question is tame. Thus Belyayev's experiment also demonstrated that animals' genetic disposition to human-friendliness could be associated with certain physical characteristics.

But the results of the experiment still left open the possibility that the tameness of the selectively bred foxes was a learned characteristic copied from the mother. In order to investigate this matter, Belyayev – who is no longer alive but whose laboratory continues to breed foxes – repeated the experiment at a later date with rats, who have a considerably shorter generation interval, so that inheritance processes are easier to observe. Here too, the aggressive rats were separated from the tame ones, only this time the most aggressive of each new litter were crossed with each other as well as the tamest. This resulted in two completely different rat populations, capable of inspiring delight or terror in today's visitors to the Siberian laboratory. The tame ones are cuddly and trusting, while the others fling themselves against the bars of their cages and snarl at visitors the moment they enter the room; if it wasn't for the barrier between them, they would probably go for the jugular. Interestingly, however, the rats' behaviour doesn't change with socialization: tame young rats who are given to an aggressive mother after birth remain tame, and vice versa. The traits of tameness and aggression must therefore be inherited, not acquired.

In 2009, a team from the MPI EVA analysed the genomes of various rat strains from the Russian laboratory and backed up the findings with DNA analyses. These showed that the most spectacular domestication results, namely the genetic changes, were achieved within the first fifteen generations; after about fifty generations the genetic characteristics of the aggressive or tame rats were more or less fixed. The human-friendly animals were those whose natural temperament had been bred out of them: their tameness was thus a mutation inherited from their ancestors. But which genetic loci in the rats were responsible for it was not known at this stage.

Then in 2018 American, Russian, and Chinese researchers turned their attention to the fox genomes from Belyayev's breeding programme and identified genome segments that appeared to be responsible for tameness. Among other discoveries, they found a variant of the gene SorCS1, which occurs with striking frequency in tame foxes but hardly ever in 'normal' or particularly aggressive ones. Until then, SorCS1 had not been linked to social behaviour but was associated, particularly in humans, with a genetic disposition to autism and Alzheimer's. On the other hand, it had been demonstrated earlier in mice that this genetic locus also plays a role in signal transmission within the cells. It is possible, then, that tame animals with a particular SorCS2 variant receive fewer stress signals when encountering humans.

Another finding worth mentioning, because it is surprising, was the existence of a connection between the genetic variant found in the particularly aggressive foxes and Williams syndrome in humans. Rather than higher levels of aggression, however, people with this condition exhibit – among other characteristics – a particular openness and friendliness towards strangers, as well as above-average musicality. But this syndrome arises from the interaction of multiple genetic changes and not just from the single variant detected in the aggressive foxes.

An acquired mistrust

The fact that an animal's potential for domestication appears to be determined by its DNA also explains why domestication has never succeeded in Africa, in marked contrast to Eurasia, where a whole range of farm animals were later bred. Hunter-gatherers in Africa failed to breed any kind of dog, nor was any domestic animal ever bred thereafter from a specimen of the continent's rich fauna – be it zebra, gnu, or one of the many wildcats. Since mutations 'rain down' uniformly across species, we can rule out the possibility that the DNA changes that produce greater human-friendliness stopped short of Africa. But it is quite clear that these mutations disappeared over time, through hundreds of thousands of years of coevolution with humans.

In Africa, because humans engaged in hunting from *Homo erectus* onwards, tame animals didn't stand a chance of survival; by contrast, those in whom the greatest fear of humans was genetically rooted and that therefore kept their distance were positively selected. By the time evolution had equipped modern humans with the intellectual resources to domesticate animals, the necessary mutations had probably long been eradicated from Africa's animals. It wasn't until their arrival in Eurasia – whose rich wildlife had hitherto had only the occasional brush with *Homo erectus*, Neandertals, and Denisovans – that modern humans encountered animals with domesticable genes. And later on they succeeded in taming not just the wolf but even species such as aurochs and water buffalo.

The genetic make-up of African animals could therefore have been one of the major reasons why the domestication of the wolf (and, for that matter, animal husbandry) began outside Africa – even if this was the result of a combination of factors rather than a single cause.[2] Indeed, it was impossible for hunter-gatherers in Africa to breed livestock, as the local fauna there knew only too well what deadly power lay within this unprepossessing biped. The world that eventually opened up to our ancestors beyond the waters of the Arabian Sea offered them a largely unresistant hunting ground.

But even this endless expanse was not enough to satisfy the incomers' insatiable desire for meat. Every piece of land they conquered brought more humans, and hence an even greater hunger for land, until the limits of the earth were reached. By the end of the global hunter-gatherer era, the various populations lived sometimes just kilometres apart and threatened to get in each other's way in the hunting fields. The end of the Ice Age ushered in a new wave of human settlement in the northern hemisphere, while conflicts over the most productive territories intensified. Post-Ice Age humans used their bows and arrows not just for hunting but also against each other, as we can tell from the holes in excavated skulls, along with other signs of combat found on skeletons from this period.

The foraging society was already on its way out by then. The centre of the world shifted to a narrow strip: the subtropical climate north of the Equator. There the foundations were laid for emerging world empires that were to last millennia. Most of humanity had passed the hunter-gatherer stage. Now it began to resist the limitations of nature by becoming a force of nature itself.

7

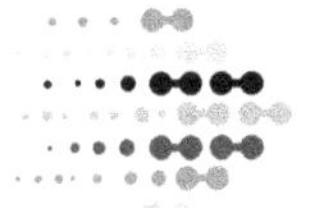

Elites

Some 15,000 years ago the hunter-gatherers begin to settle, and the end of the Ice Age ushers in the agricultural era. A bumpy road, with no way back. The new revolution sweeps the globe. Humans and beasts continue to multiply.

The first bakers

The only place where the humans of 15,000 years ago encountered untapped nature was the far east of the Eurasian continent. Via the Bering Strait – at that time uncovered by sea or ice – they colonized the New World, reaching Tierra del Fuego in the far south and establishing the Clovis culture in the north, all in record time. In America, again, they found an astonishing array of wildlife, which they annihilated with the same efficiency as on all previous occasions. Casualties this time included the giant sloth, which weighed up to three tons and measured six metres head to tail; the glyptodon, a giant two-ton relative of the armadillo; and the 'terror bird', a creature that resembled a flightless dinosaur and reached a height of up to three and a half metres. The human incomers brought their domesticated dogs with them across the strait: these 'indigenous' canines, which roamed the Aztec and Maya empires, for example, are now virtually extinct; nearly all dogs in America today can be traced back genetically to European breeds that came over from the Old Continent with the settlers.

With the Stone Age conquest of America and the extermination of the local megafauna, humans had now achieved dominance over the largest chunks of the global landmass. The next nucleus of human progress

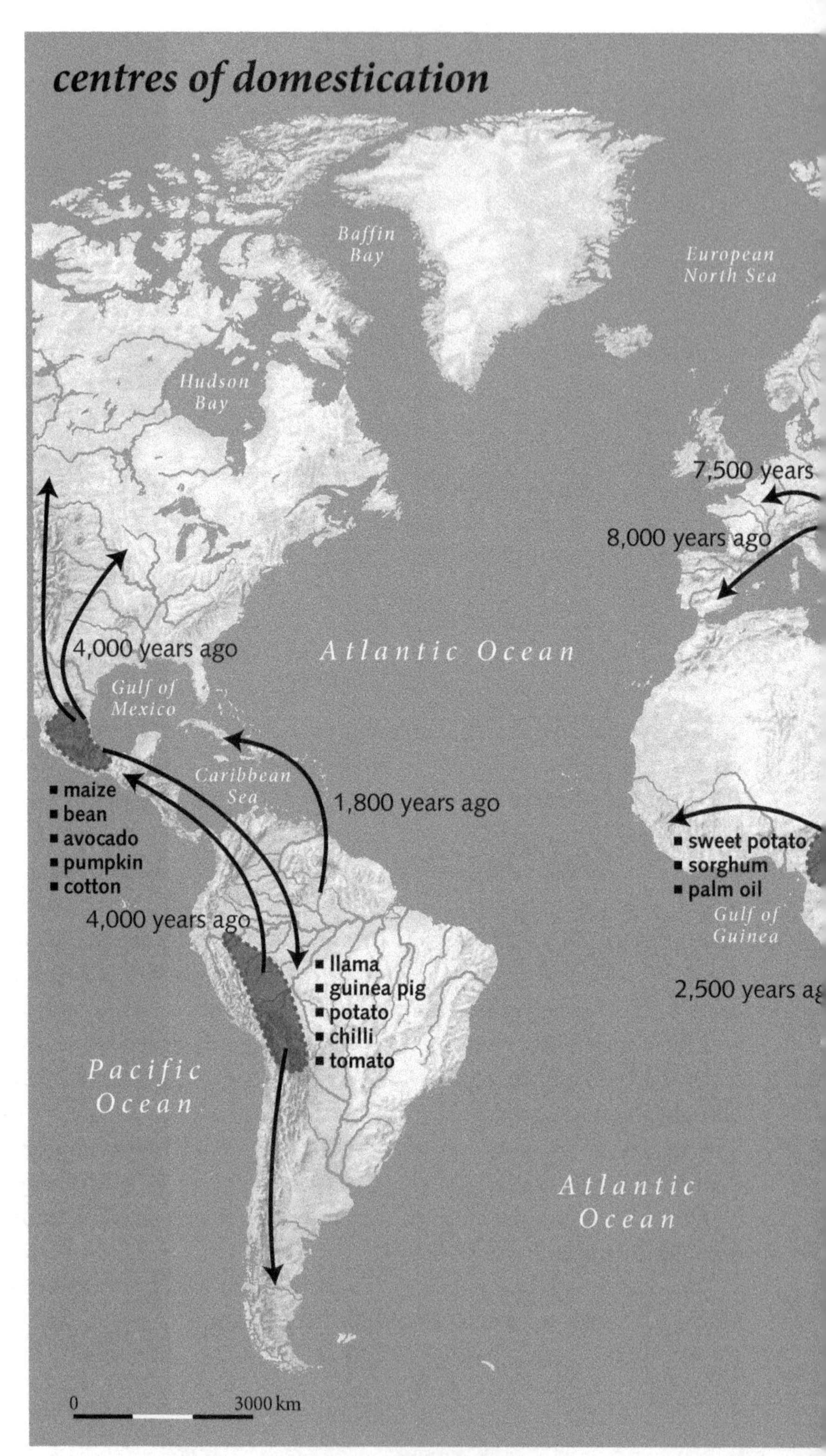

Figure 24 Centres of domestication.

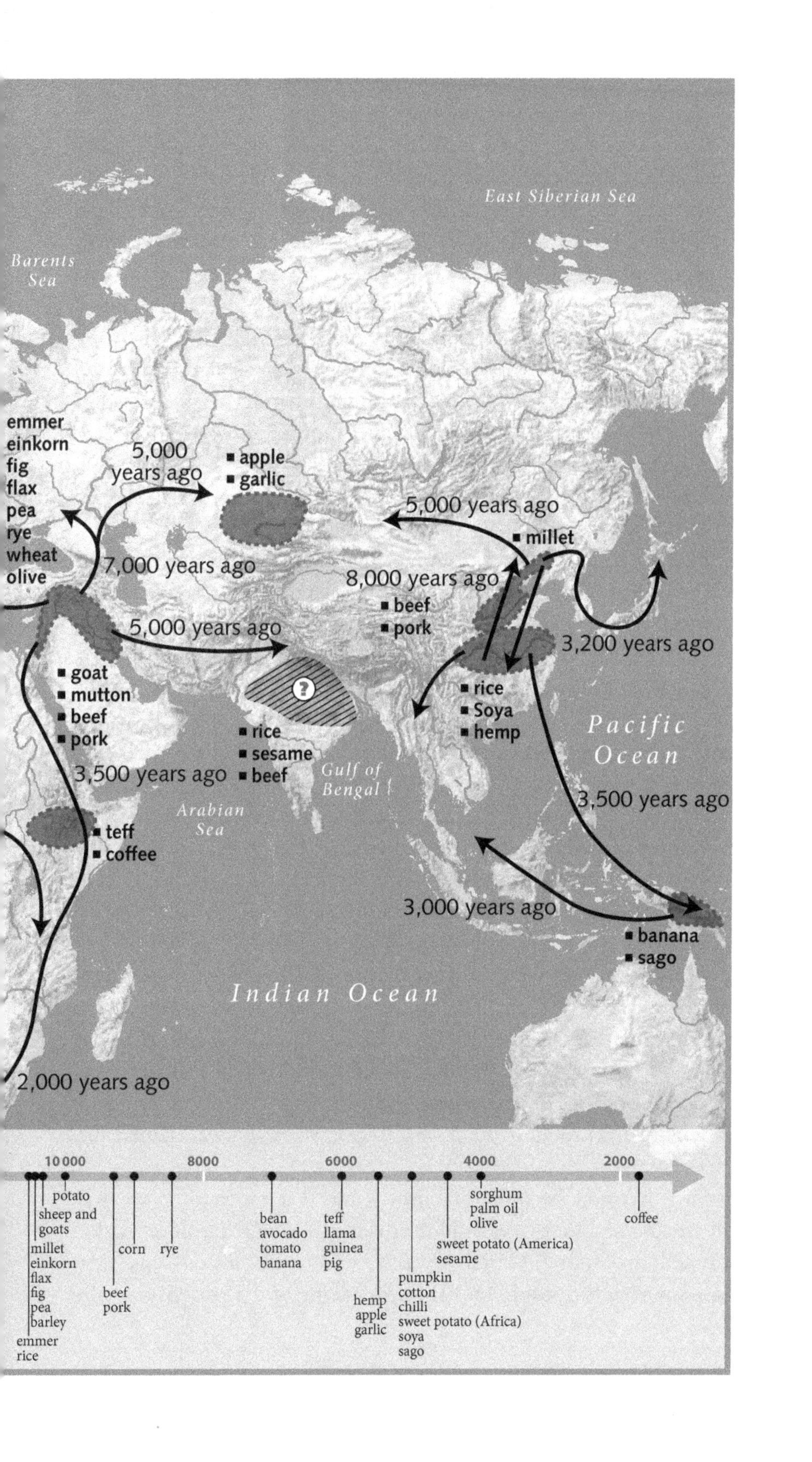
East Siberian Sea
Barents Sea
emmer
einkorn
fig
flax
pea
rye
wheat
olive
5,000 years ago
apple
garlic
7,000 years ago
5,000 years ago
millet
8,000 years ago
beef
pork
5,000 years ago
3,200 years ago
goat
mutton
beef
pork
rice
Soya
hemp
Pacific Ocean
rice
sesame
beef
3,500 years ago
Gulf of Bengal
Arabian Sea
3,500 years ago
teff
coffee
3,000 years ago
banana
sago
Indian Ocean
2,000 years ago
10 000
8000
6000
4000
2000
potato
sheep and goats
millet
einkorn
flax
fig
pea
barley
emmer
rice
corn
rye
beef
pork
bean
avocado
tomato
banana
teff
llama
guinea pig
hemp
apple
garlic
pumpkin
cotton
chilli
sweet potato (Africa)
soya
sago
sweet potato (America)
sesame
sorghum
palm oil
olive
coffee

emerged 15,000 years ago, south of the Caucasus, in the Middle East, and possibly also in North Africa. While hunter-gatherers throughout the rest of the world continued to live as they had always done, here the Natufians began to settle in dwellings in the southern Levant, that is, the region of present-day Israel, Lebanon, and Jordan. Although they still practised hunting and gathering, they no longer moved from place to place, like traditional foragers. The receding Ice Age gave way to a blossoming of growth in the Fertile Crescent, a sickle-shaped area that stretched from the southern Levant through Anatolia to present-day Iran. Alongside their traditional food sources, humans now had access to widely sprouting wild grains, which our ancestors were already using to make a kind of bread, and perhaps even beer – when they could gather enough to do so.

This practice had little in common with agriculture, as nothing was planted, or at least not deliberately. It was still a while before some bright spark noticed that spat-out or discarded grains brought forth new cereal plants. The earliest indications of an agricultural lifestyle were found, among other sites, at Göbekli Tepe in Anatolia and date back nearly 11,000 years.

A similar development occurred almost 4,000 kilometres further west, in present-day Morocco. From Grotte des Pigeons near Taforalt in the northeast of the country, we have the oldest sequenced genomes of African hunter-gatherers so far: they go back 15,000 years. To date, the cave's metre-thick rock strata have yielded the remains of around thirty-five individuals – further evidence of a long and dense colonization of Northeast Africa (remember that the 300,000-year-old skull of Jebel Irhoud came from the same area). Because of the similarity between this culture and that of European hunter-gatherers, archaeologists long assumed that the two were genetically related – in other words, that the Europeans must have spread southwards from the Iberian Peninsula via the Strait of Gibraltar. For this reason, the Stone Age culture that characterized North Africa roughly 20,000 to 10,000 years ago was named the 'Iberomaurusian'.

This term has now become established, but it is misleading; what is more, it smacks of a long-prevailing Eurocentrism in archaeology. As far as we can tell, the Strait of Gibraltar was never actually crossed – in either direction – during the Stone Age. The sequencing of the 15,000-year-old genome from Grotte des Pigeons showed no European traces. Instead, roughly a half of the DNA from these individuals can be traced back to a deeply rooted African lineage. The origin of the other half is not entirely clear. But we do know that these are the same genetic components that the Natufians carried.

The Iberomaurusians[1] and the Natufians must therefore have been genetically linked. There were certainly cultural similarities between the two populations: the Iberomaurusians, too, showed the first signs of a settled existence about 15,000 years ago; they, too, produced flour, and probably used it to make bread. Their flour wasn't made from grains though, but from wild pistachios and almonds – along with acorns, of which there are endless varieties at southern latitudes; these contain no bitter compounds and are therefore edible, like nuts. Other archaeological parallels with the Natufians have also been found, for example in the production of jewellery and tools.

Mysteries of the Orient

Although the artefacts of Iberomaurusians are a few thousand years older than those of Natufians, the genetic bridge between the two populations cannot be explained in terms of an expansion of humans from Morocco into the Fertile Crescent – otherwise the Natufians would also have carried the deeply rooted African DNA of Iberomaurusians. But this is not the case. On the other hand, all Iberomaurusians carried Natufian DNA, which made up the other half of their genome, alongside the African half. The obvious explanation that the Natufians spread their genes westwards is unlikely, given the timeline: after all, their culture didn't develop until long after the beginning of Iberomaurusian culture, which means that they cannot have been the ancestors of the Iberomaurusian people. The genetic bridgehead between the two cultures must have been located elsewhere.

The most likely scenario linking the genetic and archaeological coordinates has the ancestors of the Natufians – who later advanced westwards and eastwards – living somewhere between Morocco and Israel. The Natufians who went east ended up in the Fertile Crescent, which was virtually devoid of humans in the wake of the Ice Age, and they spread without mixing with any other population. In the west, by contrast, they came into contact with African hunter-gatherers – perhaps the people of the Aterian stone tool industry, a forager culture that predated the Iberomaurusian and covered present-day Morocco and Algeria, among other areas.

All this would suggest that the mysterious Basal Eurasians – the ghost population from which Natufians derive half their ancestry and whose DNA subsequently found its way to Europe through the expansion of agriculture – may also have originated from somewhere in North Africa, such as the highly fertile land of Egypt. Although there are no skeletons here from that time, this is readily explained by the virtual impossibility of finding archaeological traces under the metre-thick layers of sediment in the Nile Delta. But that doesn't bring us any closer to solving the mystery of where the Basal Eurasians lived who later merged with the mixed population of modern humans and Neandertals to become Natufians. It could still be anywhere from east of Morocco all the way to Iran, though it is most likely to have been somewhere between Egypt and the Arabian Peninsula. The same goes for the Natufians themselves, whose roots may likewise lie in this very broadly defined region, either within Africa or outside it.

Palefaces

As the Ice Age waned, so agriculture began to flourish. Naturally, this wasn't something that humans developed overnight, but through a gradual process of trial and error. Rising temperatures made it possible to cultivate multiple varieties of grain, including the early forms of wheat and barley. Agricultural life was by no means superior to life in hunter-gatherer societies, which of course continued to exist alongside it in the Middle East. On the contrary, settlement demanded the labour of everyone in the community; at the same time there was the constant risk of storms and droughts, which endangered not only the current season's harvest but also the seeds of the next one – an existential threat that hung henceforward like a Damocles sword over every farmer's life. Furthermore, these early attempts resulted almost inevitably in malnourishment. The first farmers didn't keep cattle, and even when they began to do so a few centuries later they used them mainly for milk production: as a rule, the animals were slaughtered only when they were no longer productive. This left people short of vital nutrients.

Nor could the hunger for meat that had evolved over millennia be satisfied by switching back to hunting: there wasn't enough time, and the necessary know-how had been lost anyway after a few generations of farming. Humans lost their connection to nature. But agriculture obviously had its advantages too, or else it wouldn't have gone on to conquer the world. With hard work, one could ensure a sufficient and reasonably predictable food supply without the burden of having to keep moving all the time: at last, people had a secure roof over their heads.

Plannability, a place to live, a reliable income: these things are still regarded by most people today as the ideal basis for starting a family. And that's just what Middle Eastern farmers did. They multiplied continuously, spilling into new fields, new settlements, and new pastures. From the Fertile Crescent they spread out in all directions, changing the genetic structure not just of large parts of Eurasia, but of Africa too. Later on agriculture established itself entirely independently in other regions of the world as well, supplanting forager cultures everywhere as they struggled to compete with the promised benefits of this brand new lifestyle.

The signature technology of the Neolithic period was pottery, which served to make vessels for cooking, eating, drinking, and storage. Storage was much in demand, as the new environmental conditions, with sufficient rain and plenty of sun, tended to ensure rich harvests. What's more, the local fauna positively invited domestication: the wild forms of the sheep, goats, cows, and pigs that began to be artificially bred some 10,000 years ago were all concentrated in the Fertile Crescent. Equipped with their Neolithic tools and with knowledge of cereal cultivation, farmers – albeit less mobile than people in the hunter-gatherer societies of the rest of the world – were streets ahead when it came to subjugating nature.

Figure 25 The hunter-gatherers of Europe © Tom Björklund

The hunter-gatherers of Europe probably had a much darker skin than the farming folk, who arrived later on the scene. How the pigmentation gene inherited from the Neandertals influenced our ancestors' complexion is, however, open to question.

Europe was the first to feel the full force of that aspiration, after the Neolithics spread to the continent from Anatolia about 8,000 years ago. They began by colonizing the Balkans, then travelled along the Danube and the Mediterranean to the most fertile regions of central, western and eastern Europe before advancing to Scandinavia and the British Isles, roughly 6,000 years ago. This expansion also led to the loss of the hunter-gatherers' original skin colour, which was permanently replaced by the paler complexion of the incomers (Figures 25 and 26). The lightening effect was due to their agrarian diet, which was probably very low on meat, and began with the first Neolithic lifestyles in the Middle East, before eventually spreading and intensifying across the entire northern hemisphere. Light skin came about through multiple mutations, which then presumably became established as the only way of ensuring that enough sunlight was absorbed through the skin to ensure a sufficient supply of vitamin D. This was essential for a long life – and thus for producing numerous offspring, to whom the mutations could be passed on. The hunter-gatherers didn't need this workaround, as they obtained enough vitamin D already through their copious meat and fish consumption.

In sub-Saharan Africa, on the other hand – as in Central and South America – a pale skin did not confer any selective advantage, even after the advent of agriculture: because of the intense sun in these regions, enough light can be absorbed to keep the body well supplied with vitamin D without the need for such mutations. Consequently, the original skin colour of our ancestors was largely retained along the Equator, from Africa through Southeast Asia to America, long after agrarian lifestyles took hold there, or even began to emerge.

Figure 26 The transition to an agrarian society © Tom Björklund

The transition from a forager society to an agrarian one took a great deal of labour and was often accompanied by malnutrition. Despite this, the global advance of agriculture proved unstoppable.

The dominance of the newcomers

The farming folk advanced not only northwestwards, to Europe, but also to Africa, and right down to the south of that vast continent. In the east the Neolithic revolution spread to the Indian subcontinent, and in the north, via the Caucasus, to the European steppe. Archaeologists have long known about this expansion, but the genetic data obtained in recent years have helped to build a much fuller picture of how it happened.

Up until then, the jury was out – particularly concerning Europe – on the question of whether it was the Neolithic culture that spread or the Neolithic people. In other words, did the hunter-gatherers of Europe absorb the idea of a settled lifestyle as a cultural import or were they displaced by their neighbours? The genetic data couldn't be clearer on this point: Europe's foragers were dominated and pushed aside – and within just a few centuries. The inhabitants of Europe before the Neolithic expansion show clear genetic differences from the incoming farmers, whose bones were found to contain the same DNA as those of the people who lived in Anatolia at that time.

Just as in the case of the arrival of our ancestors, who took over the Neandertals' land in Europe and the Denisovans' in Asia, here too we don't know how aggressive the expansion of the first agrarians was. Their lifestyle and the resulting possibility of producing more offspring certainly suggest an increasing numerical superiority over the hunter-gatherers. That doesn't necessarily mean that the displacement came about through destruction, though. It may be that the hunter-gatherers retreated to less fertile areas, notably in northern Europe. The genetic data would support this: the forager DNA is most prevalent in present-day Scandinavian and Baltic people, accounting for roughly the same proportion as the agrarian genes. In the fertile regions south of Central Germany and France, by contrast, the hunter-gatherer components decrease the closer we get to the starting point of the Neolithic revolution.[2] The phase of interbreeding between the existing population of Europe and the newcomers lasted almost 2,000 years: before that, the native foragers and the descendants of the Anatolian immigrants seem to have done their best to keep out of each other's way.

Strictly speaking, there were two Neolithic revolutions in the Middle East that took place simultaneously yet independently of each other – which is further proof of how fertile this region became after the end of the Ice Age. In addition to the agricultural revolution in Anatolia and the Levant, there was also the Iranian Neolithic. This, as the name suggests, had its origins in what is today Iran and was initiated by people who, though direct neighbours of Anatolians, were genetically very different from them. The two populations must have been previously separated for tens of thousands of years by an insurmountable barrier, such as ice-covered mountain ranges or deserts – the only explanation for such genetic differences. The genes of the Iranian farmers can still be detected in the genomes of modern humans as far as northern India; when and how they got there has yet to be conclusively established.

The moot point is whether the Neolithic developed autonomously on the Indian subcontinent or it, too, spread there from Iran via present-day Afghanistan and Pakistan. The latest analyses of ancient DNA certainly suggest an autonomous development. There is, however, no doubt that the Iranian genes spread via the Caucasus to the Eurasian steppe at that time. In this region, the incomers bred with the local hunter-gatherers, and later on the Yamnaya culture emerged within this mixed population.[3]

There are good reasons why the farmers dispersed primarily along similar latitudes rather than advancing deep into northern Siberia, like the hunter-gatherers before them. For one thing, the climatic conditions in the regions where the Neolithic took hold were not too different from those of its region of origin. Central Germany is a mere thousand kilometres north of Anatolia, and the Caucasus is even closer. Latitude was, along with soil composition, one of the key factors governing the success of Neolithic expansions: after all, migrant farmers could settle in their new home only if they could make the domesticated plants they brought with them grow there too. If the temperature, the soil, the rainfall, or the seasonal rhythms were substantially different, the imported grain varieties had little chance of survival. In Europe this was not the case: the imported precursors of today's wheat and barley flourished in the loess soils formed there during the Ice Age. The same went for the Iranian Neolithic between the Caucasus and – if the farmers got that far – northern India.

The Ancient Egyptians: A closed shop

Large parts of Eurasia thus offered an ideal living for the expansionist farmers of the Middle East. Unsurprisingly, the latter set their sights on Africa too, the southern Mediterranean coast in particular. Thanks to the greening of the Sahara about 10,000 years ago, the conditions there were now ripe for agriculture and the land opened up to migrants all the way to Morocco. Consequently, the DNA in the Fertile Crescent today represents a genetic bridge that connects Europeans with North Africans.

How strong the genetic influence of the Neolithic migrants was in this region is, however, very hard to judge. After all, the roots of the Natufians, whose DNA was present in the farmers who came to North Africa, may have lain in the very same region – at least, archaeological traces of them have also been found on the Sinai Peninsula. Furthermore, the lack of sequenceable ancient DNA makes it difficult to determine which components came from the Middle East to North Africa some 7,000 years ago and which ones were already there. As for whether Neolithic people displaced the long-established hunter-gatherers, as they did in Europe, that too is a question that cannot be resolved with the scant genetic data.

What is certain is that the Nile Delta (Figure 27) in Egypt became the undisputed epicentre of North African agriculture. Despite being less than the size of Belgium, it currently has six times its population: around 60 million Egyptians live in this extremely fertile valley, which makes it one of the most densely populated regions of the world. The growing conditions here have always been ideal and were probably even better than in Europe at the beginning of the Neolithic period.

The oldest DNA from Egypt comes from the country's most famous cadavers: mummies. In 2017 mitrochondrial DNA from almost a hundred different Egyptian specimens was sequenced, and so was the genome from three of them. The specimens under study belonged to people who lived sometime between 3,400 and 1,500 years ago and were all buried in what is now the central Egyptian town of Abusir el-Meleq. They were perfectly preserved (as if to order), thanks to the ancient Egyptian technique of mummification, which continued under Roman rule and was widely practised, even on dogs, cats, and other animals. The long period during which the sequenced individuals lived saw the high

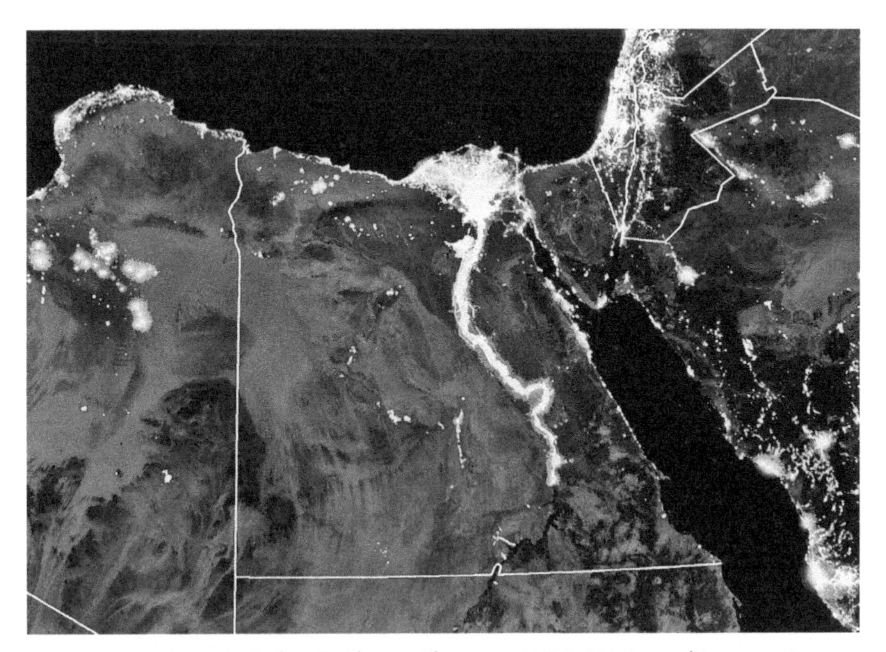

Figure 27 The Nile Delta © Planet Observer / UIG / Universal Images Group North America LLC / Alamy Stock Photo

At the beginning of the Neolithic period, the Nile Delta was one of the most fertile places on earth. To this day, there are few regions as densely populated as northern Egypt. The bright lights of the delta can even be seen from space.

point of Egyptian civilization – what we call the New Kingdom – as well as various other periods of foreign domination, be it by Nubians, Assyrians, Greeks, or Romans. Surprisingly, however, no genetic admixtures from other regions – including sub-Saharan Africa – were found in any of the mummies. So it appears that before the heyday of ancient Egypt there was no measurable gene migration from the south of the continent.

The situation is very different for the people who live in the country today: Egyptians carry around a fifth of sub-Saharan DNA. This gene flow, which appears to have begun in late antiquity at the earliest, could have been brought about by increased migration or trade with the south. Another, equally possible cause is the massive slave trade operated by Arabs, who came to Egypt in the seventh century: an estimated six to seven million people were trafficked from South to North Africa at the time.

Learner livestock farmers

While the Sahara blocked any gene flow between the northern and the southern parts of East Africa up until the historical age and beyond, the Nile eventually became a corridor. By comparison with the African hunter-gatherers, who failed to conquer this barren region because of the dearth of food sources in it, the farming folk of the Levant had a much better tool in their armoury: animal husbandry or, more precisely, 'pastoralism'. This is a very specific form of livestock farming, whereby the herds are constantly made to move on when the natural grass and shrublands become depleted. In this way the people from the Middle East began to fan out along the Nile and the Great Rift Valley some 4,000 years ago, advancing with their sheep and goats via Sudan to Ethiopia. Unlike in the north, however, the incomers didn't displace the local hunter-gatherer populations: only a fifth of the genetic make-up of those who lived in this region immediately after dispersal can be traced back to the pastoralists.

In contrast to farmers in Europe and Asia, who relied mainly on the grains cultivated in Anatolia, the pastoralists lacked this economic pillar: they were unable to export successfully to Africa the plants domesticated by their ancestors because of the many different latitudes and climate zones they had to negotiate as they dispersed southwards. Their mainstay was the livestock they brought with them. Thus, rather than supplanting the hunter-gatherer model, pastoralism – which depends similarly on a degree of mobility – initially subsisted alongside it, on equal terms. The only region where the newcomers were able to establish any kind of agriculture was the Ethiopian Highlands, where they succeeded in domesticating teff, a type of millet now popular worldwide as a gluten-free alternative grain.

The pastoralists spread right down to the south of Africa, and their genetic signature is still found today in many peoples of this region. One example is the Khoisan, who live mainly in Namibia, South Africa, Botswana, and Angola. Some ancestors of the Khoisan split off from the modern human lineage more than 200,000 years ago – earlier than any other still extant population. Then, 2,000 years ago, they began to absorb the incoming pastoralists from the north, who account for about 10 per cent of present-day Khoisan ancestry. Members of the various Khoisan groups show clear phenotypical differences from their neighbours, such as a generally paler skin. This is not necessarily an import from the Middle East, though. The darker skin colour of most other Southern Africans is rather to be traced back to another Neolithic revolution – the Bantu expansion.

The Bantu expansion began about 4,500 years ago in West Africa, in the region now occupied by Nigeria and Cameroon. Unlike the Neolithics from the Middle East, the Bantu had the advantage of being able to bring their root crop-based agriculture with them as they dispersed along the same latitude – probably across the Ugandan plateau and southern Sudan towards East Africa – about 4,000 to 3,000 years ago. There they encountered the pastoralists who had previously migrated from the Middle East and gratefully adopted their idea of crossing the savannah with domesticated, milk-producing sources of meat, before finally advancing southwards as full-fledged livestock and arable farmers.

Boasting 200 million speakers and some 500 varieties, the Bantu language family now dominates the entire southern half of Africa, whereas the Khoisan languages, though still spoken in the southwest, have become marginalized. Today, at any rate, the genes introduced from West Africa by the Bantu are distinctly preponderant in the southern half of the continent, so the Bantu farmers must have achieved the same kind of cultural dominance as the Anatolians had done in Europe. This migration was male-driven, as we know from the genomes of the people who live in the region today. The gender-specific DNA segments are unequivocal: while the paternally inherited Y chromosomes came to the region almost exclusively through the Bantu expansion, the maternally inherited mitochondrial DNA derived largely from the female foragers who were already living there at the time. This means that the incoming male farmers fathered children with the local women, while the male hunter-gatherers appear to have done so rarely if at all. What role violence played in this phenomenon cannot be deduced from the genetic data, but the local men are unlikely to have surrendered their right to offspring voluntarily.

One blank spot on the archaeological map is the Congo region – a fact that has much to do with the complicated political situation there, aggravated as it is by military conflicts. Getting access to archaeological samples from the region has proved so far virtually impossible; add to this the poor conditions that tropical humidity offers for the preservation of bone DNA. Nevertheless, a small number of ancient DNA samples have meanwhile been sequenced, and they suggest that hunter-gatherers in this largely rainforest-covered land were not displaced by farmers: the tribes that continue to forage in the rainforest today are indeed genetically very different from members of the Bantu populations. The same is true of the Hazda – a hunter-gatherer group of some thousand souls in northern Tanzania – who still exhibit a genetic mix that stretches back to the first Africans. There is virtually nowhere else within the geographic range of the Bantu where the DNA of the ancient African hunter-gatherers still survives.

At the same time, the genes of the pastoralists from the Middle East that were still clearly present in the East Africans of 4,000 years ago have also been much diluted over millennia, their signature being almost completely overwritten by the encroaching Bantu. Paradoxically, the Khoisan in southern Africa – who, we remember, go back partly to the pastoralists from the Middle East – are thus genetically much closer to the Anatolians than are the East Africans, who owe most of their genetic make-up to the Bantu expansion. Thus, long before the first Europeans colonized southern Africa in the seventeenth century, a genetic trail had already led there from outside Africa; and it is a trail that connects with the deepest human lineage on the planet.

Eternal China

Several hundreds of thousands of years of human and animal coevolution had made Africa a far less favourable place for domesticating animals, so that the inhabitants of that continent entered the Neolithic race with a significant locational disadvantage. Circumstances were very different in the Far East, where agriculture – despite having started a tad later than in the Fertile Crescent – soon took off with a vengeance. There too, in what is now China, the Neolithic revolution was largely built on cereal crops to begin with. Like the area between the Euphrates and Tigris in the Middle East, the land between the Yellow River (Huang He) in northern China and the Long River (Yangtze) further south remains one of the most fertile regions of the world. The Neolithic revolution began along the Yellow River around 10,000 years ago, with the domestication of Asian millet. Further south, people began to cultivate rice around the same time, and in a few centuries dense settlements sprang up along both rivers.

And there were further parallels with the development of agriculture in the Fertile Crescent. Just as, in the Middle East, two different hunter-gatherer populations acquired farming know-how in the immediate vicinity of each other at the same time, the genomes of the people living along the Huang He and the Yangtze at the beginning of the Chinese Neolithic are also distinct from each other, if not quite as obviously. Furthermore, we find a similar genetic continuity with the foragers who earlier inhabited both regions. This means that the Neolithic was a domestic invention in China too.

Neither the farmers in the north nor those in the south seem to have gained the upper hand, as there was, over time, an even flow of genes in both directions that did not cause any genetic displacement. Both populations evidently interacted peacefully, and very likely also traded with each other; archaeological finds show at least that some time later millet spread from north to south, and rice in the opposite direction.[4]

This long phase of near vegetarianism, which here too was certainly attended by malnutrition, was later followed by a period of animal domestication. There were no sheep or goats in East Asia but plenty of water buffalos and pigs, whose purpose-bred variants became part of the Chinese farmer's inventory at least 9,000 years ago. As for today's most widely devoured domestic animal, the chicken, that, too, has its origin in Southeast Asia, being a descendant of the red jungle fowl, which is still found in this region today.

Like the Mediterranean, the fertile land of China – that is, the south-eastern part of the republic we know today, where the vast majority of Chinese people live – offered ideal conditions for the development of a flourishing culture. And so it proved to be. After the emergence of a whole series of highly complex cultures around the Mediterranean in the first millennium BC, China followed suit about 2,300 years ago. But, while the ancient Greek, Egyptian, and Roman city-states and empires all eventually collapsed, the Chinese empire, for all its many cultural upheavals, proved astonishingly stable over the millennia – from the first ruling dynasty, some 2200 years ago, to the last emperor in the early twentieth century. By way of comparison, imagine that the Roman Empire (which started in the first century AD, more than 2,000 years after the Chinese) or the kingdom of the pharaohs still existed today. This stability is also reflected in the genes of contemporary Han Chinese, who account for more than 90 per cent of the population. They can be traced back in large measure to the country's first Neolithics, as well as to early inhabitants of central and eastern China.

This genetic continuity has much to do with China's natural boundaries, which made it largely resistant to waves of migration from the rest of Eurasia. The Gobi Desert and the Himalayas in the west and south of the country, for example, present virtually insurmountable obstacles. These barriers served to insulate China while slowing the progress of the Chinese Neolithic westwards. The only remaining route – the northern steppe – was too cold. The gateway to the world – which the East Asian Neolithics later forced open in order to spread their DNA across half the planet, during what was possibly the boldest migration movement in human history – lay in the east, on the Pacific coast.

Trade-happy Americans

As we have seen, the Neolithic revolutions of Africa and Eurasia were characterized by constant waves of migration and displacement, invariably initiated by the dominant farmers. Thus the Anatolians shaped the genetics of Europe, the Middle East, and parts of sub-Saharan Africa; the Iranian Neolithics advanced eastwards, perhaps as far as India and into the Asian steppe; Bantu genes now dominate almost the entire southern part of Africa; and the Chinese farmers spread across the entire fertile territory of their vast empire. In America, which had been isolated from the rest of the world since the end of the Ice Age, the Neolithic revolution took a very different form. There it was not associated with any genetic shifts that resulted from the dominance of incoming farmers. On the contrary, the modern descendants of the original North and

Figure 28 Comparison between teosinte, a maize–teosinte hybrid, and modern maize © John Doebley / Wikimedia Commons / CC-BY-SA-3.0

Comparison between teosinte (top), a maize–teosinte hybrid (centre), and modern maize (bottom). The cultivation of maize from teosinte is thought to have begun more than 6,000 years ago.

South Americans are genetically similar to those who lived there as long as 12,000 years ago.

Americans' first attempts to cultivate a humble variety of wild grain – probably in the north of Mexico and in the southern part of the Rocky Mountains – are thought to have begun more than 6,000 years ago. Teosinte still exists today and is characterized by an extremely sparse appearance and just a few kernels per ear; it was from this plant that the maize we know today was bred, about 3,000 years ago (see Figure 28). After wheat, which originated in the Fertile Crescent, and rice, which was developed in China, maize is now one of the three most popular types of grain on the planet. The tomato, too – one of the most commonly cultivated vegetables, since 200 million tons of it are currently produced each year – has its origin in America, as does the potato, which has revolutionized food security in modern Europe. It is all the more astonishing, then, that the cultivation of these plants never resulted in the dominance of the American peoples who first invented them.

One logical explanation is the longitudinal orientation of North and South America, which may have hampered farmers' attempts to expand with their Neolithic equipment. Even so, both maize and crops such as potatoes, tomatoes, and a variety of pulses quickly spread throughout South America after the Neolithic revolution, which suggests that these fairly robust plant species weren't overly compromised by the different climate zones. Nor were there any major natural barriers on either continent – both of which, as we know, were traversed from north to south in record time after the arrival of hunter-gatherers. No: there must have been other reasons why the American Neolithic followed a very different script from that of Eurasia and Africa. Although we don't know where the various agricultural plants originated – and even in the case of maize we can only make an educated guess – there is much evidence to suggest that successful cultivation attempts occurred simultaneously in several parts of Central America. For the neighbouring populations, this would have created strong incentives for trade. After all, you can do much more with a combination of tomatoes and potatoes than with just tomatoes or just potatoes.

Another factor militating against the dominance of one particular population was the almost complete absence of domesticable animals. Although the North American prairie teemed with buffalo until the arrival of modern Europeans, there were no attempts to breed them, perhaps precisely because there was such a copious supply of these meat-bearing animals. The subsequently domesticated llamas and alpacas were mainly used as wool-bearing beasts of burden, and the guineapigs that were bred for consumption were clearly not reason enough to grant one group total superiority over another. As a result, even the Mayan Empire, which extended across present-day Guatemala and parts of South Mexico and rose to become a beacon of culture in the first millennium, brought scarcely any genetic shifts in America – let alone of the kind observable around the Middle East or among the Bantu in Africa. Although agriculture spread like wildfire through North and South America over the following millennia, it left the genetic structure of both continents largely untouched, apart from a small amount of gene migration from Central America to the south; only in the Caribbean were the local hunter-gatherers supplanted by the incoming farmers, giving rise to the Taíno culture.

The foragers' long goodbye

By the time the Neolithic had established itself throughout America, the whole planet was virtually covered in agricultural systems, Australia remaining a rare exception until the arrival of the Europeans. On neighbouring New Guinea, the world's second-largest island, a mixed system developed by the locals took root that was based largely on slash-and-burn agriculture and banana growing. There as in the rest of the world, hunter-gatherers have survived only in tiny enclaves if at all, and these groups look likely to disappear altogether within the next few decades. With the parallel rise of agriculturalists and the rapid increase in population[5] came the extreme expansion of a small number of animal populations on whose exploitation our progress depended. While humans and their domesticated animals accounted for around one thousandth of the mammals that inhabited the planet 10,000 years ago, that figure is now 95 per cent – an explosion that, for all the staggering population growth of our own species, is chiefly attributable to our domestic animals.

Whereas for hundreds of thousands of years the prospects of the global forager community were limited by what the earth had to offer, our agrarian ancestors managed within a fraction of that time to access seemingly inexhaustible opportunities for growth. They got better and better at optimizing plants and animals to their liking. The threat that the wilderness posed from time immemorial was contained by the force of civilization. The demand for new land increased beyond all measure. From China, humans proceeded to conquer a paradise that, soon after their arrival, became unrecognizable.

8

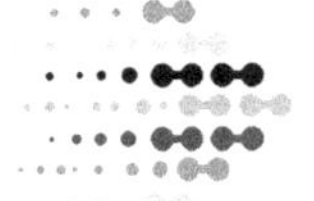

Beyond the Horizon

About 5,000 years ago we run out of space; from China and Taiwan, we take to the Indian Ocean and the Pacific, conquering one island after another and raiding their resources as we go. We risk our own existence for the sake of status symbols.

Cattle trucks on the high seas

Agriculture flourished in China just as it did in Anatolia, and there, too, the available arable land was soon outgrown. The early Chinese farmers colonized Thailand and Vietnam some 6,000 years ago, then moved on to Sumatra. The hunter-gatherers of what is known as the Hoa-Binh culture had to make way, and the DNA of people as far away as Borneo and Java still bears the stamp of the southern Chinese farmers. The same thing happened in the north, where the farmers reached the Korean Peninsula first, then crossed to the Japanese Archipelago about 3,500 years later. There they encountered a highly developed hunter-gatherer culture that, like the Natufians in the Near East many thousands of years before, already showed signs of settledness, though it's debatable whether Japan's original inhabitants already practised early forms of agriculture. What is certain is that the Jōmon people were already producing ceramics 15,000 years ago, and this makes them the world's oldest potters. In the late phase they became remarkably skilled at this craft (see Figure 30).

But against the migrants from the Chinese mainland they didn't stand a chance: once the newcomers, over the course of a few centuries, established themselves and spread across the entire Japanese Archipelago, the Jōmon culture duly disappeared. The hunter-gatherers were literally

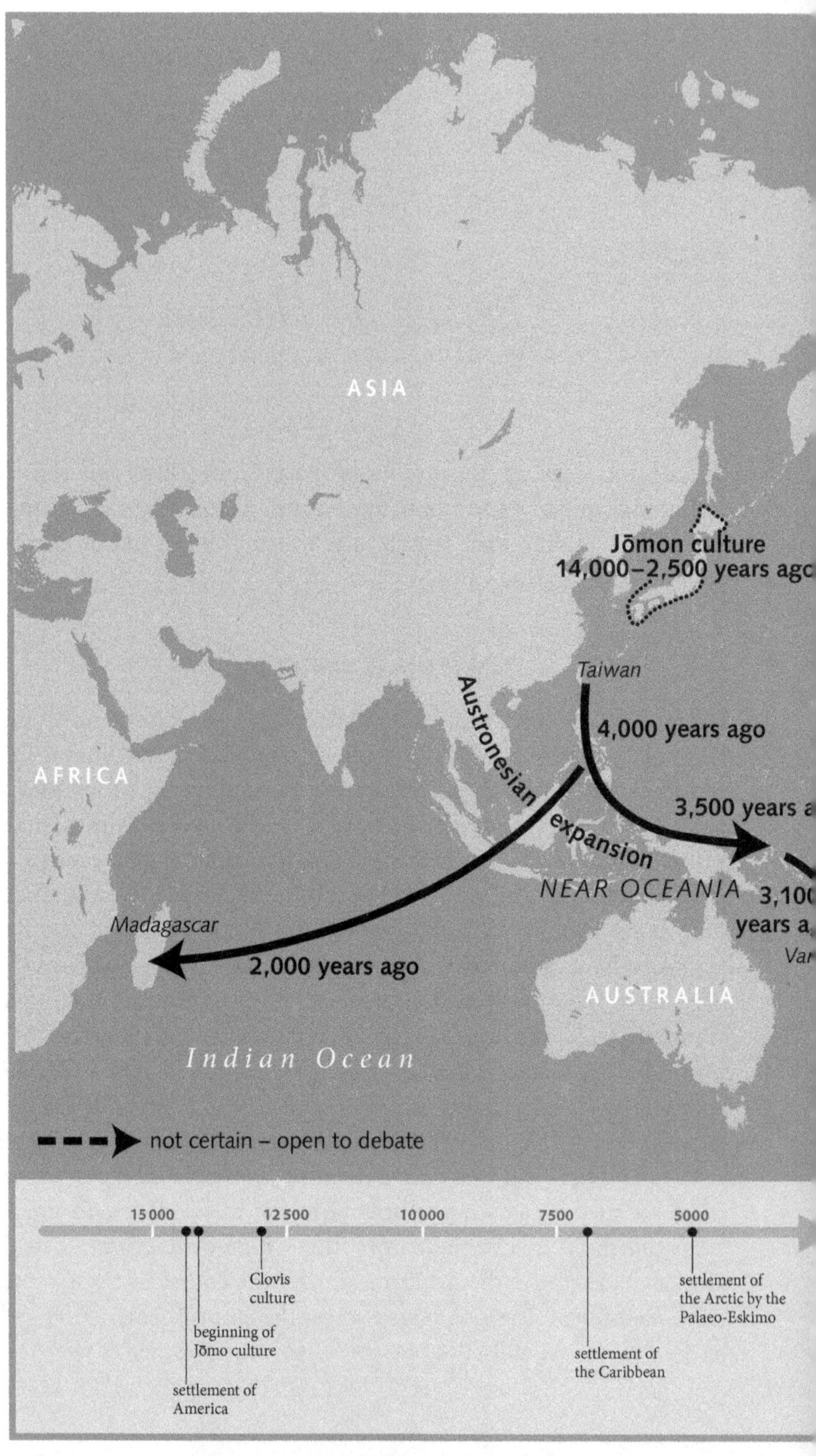

Figure 29 Beyond the horizon.

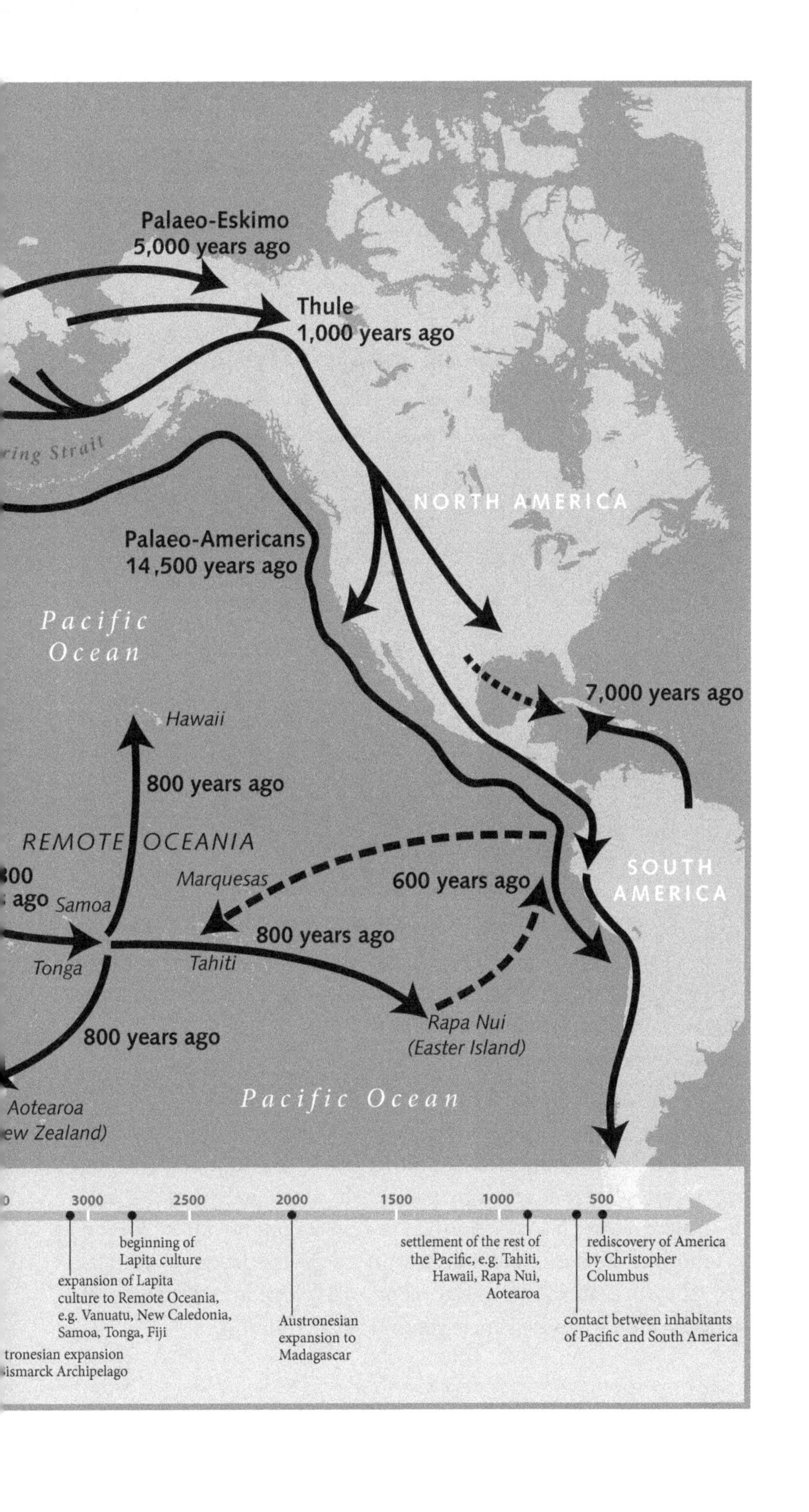
Palaeo-Eskimo
5,000 years ago
Thule
1,000 years ago
NORTH AMERICA
Palaeo-Americans
14,500 years ago
Pacific
Ocean
7,000 years ago
Hawaii
800 years ago
REMOTE OCEANIA
Marquesas
600 years ago
Samoa
SOUTH
AMERICA
800 years ago
Tonga
Tahiti
800 years ago
Rapa Nui
(Easter Island)
Aotearoa
Pacific Ocean
3000
2500
2000
1500
1000
500
beginning of
Lapita culture
expansion of Lapita
culture to Remote Oceania,
e.g. Vanuatu, New Caledonia,
Samoa, Tonga, Fiji
Austronesian
expansion to
Madagascar
settlement of the rest of
the Pacific, e.g. Tahiti,
Hawaii, Rapa Nui,
Aotearoa
rediscovery of America
by Christopher
Columbus
contact between inhabitants
of Pacific and South America

Figure 30 Jōmon pottery

By the time Chinese agriculture came to the Japanese archipelago, a highly complex hunter-gatherer culture had already developed there. The Jōmon people are among the world's earliest potters, producing richly ornamented ceramics as long as 15,000 years ago.

marginalized: the highest proportion of ancient Japanese forager DNA is found on the northernmost and southernmost of the islands, where it makes up on average a tenth in present-day Japanese people – and more than twice as much in the north. The highest proportion of Jōmon genes is found in the Ainu, the descendants of the last Japanese hunter-gatherers. They live on the northern island of Hokkaido, with many of them deriving more than half their genes from the Jōmon.

Just like Japan, the island of Taiwan to the southeast of China was part of the Asian landmass during the Ice Age. It too was inhabited by hunter-gatherers before the incoming East Asians took over the island some 5,000 years ago. But, unlike Japan – where, after the farmers' colonization, a remarkably homogeneous genetic structure imposed itself, with scarcely any interaction with the mainland – Taiwan was to become a bridgehead for the last big wave of human expansion. Over a period of several thousand years, East Asian genes kept spreading across half the globe, from the African coast at the extreme western end of the Indian Ocean to the American side of the Pacific.

Humans conquered, one by one, all the islands of the 'water hemisphere' – that is, the side of the earth that is covered by 90 per cent water – using catamaran-like boats with giant sails. These were loaded with pigs, chickens, rats, and crop plants domesticated in their native lands. In their quest for the last available pieces of useable land on the planet, the explorers travelled hundreds, and often thousands of kilometres, without knowing where their journey would take them. The genetic trail of the legendary 'Austronesian expansion' leads from Madagascar via New Zealand to Hawaii and Easter Island. For the smaller Pacific islands in particular, this human conquest meant the exploitation of all available resources. The only way the settlers could solve the problem of their finiteness was to move on to the next islands – until there were no more left to discover and the growth-driven lifestyle imploded in many places (see Figure 31).

Island hopping

The Austronesian language family today comprises 1,150 languages, spoken by some 300 million people. These include Malagasy, the language of Madagascar's Indigenous inhabitants, and the language of the Māori in New Zealand. The common origin of all Austronesian languages in Taiwanese – whose modern variety is spoken by around 80 per cent of Taiwanese people – has been established through linguistic analyses that point to a split about 5,000 years ago. Around that time the Taiwanese, who had built up a flourishing agricultural lifestyle with pigs, chickens, and rice plants, seem to have outgrown their island. It would be wrong, however, to regard the Austronesian expansion as a sudden departure led

Figure 31 Reconstruction of a boat used by the explorers of the Austronesian expansion in their conquest of the eastern Pacific

They probably used catamarans, whose cargo holds were also loaded with domestic animals and seeds with which to start new life in a foreign land.

by a group of bold adventurers who set out to sail the world in order to reach Madagascar, 9,500 kilometres away from Taiwan, or Easter Island, 18,000 kilometres to the southeast.

Rather, the conquest of the Pacific islands took the form of a kind of island hopping whereby the settlers colonized the nearest island first and – a few generations later, when there was no more land left to distribute – spilled onto the next, leaving the world of their ancestors behind. These expansions may have been embedded in fixed initiation rites. It is quite conceivable, for example, that elites were formed on the islands – elites that consisted of those whose offspring risked their lives in search of distant isles. If they failed, the family mourned a lost child; if they succeeded, the explorers would pilot a dozen or more families to a new island, where they would then go on to rule. Balanced against the risk of going down to the bottom of the ocean was the opportunity of going down in the annals of history as a legend. The Austronesian expansion was sustained by bounty hunters.

This great voyage began in the tropical island realm of Micronesia, a group of around 2,000 islands and islets, some of them tiny. This part of the western Pacific occupies approximately 7 million square kilometres, covering a width of around 4,000 kilometres east of the Philippines. To the south – that is, east of Australia – Micronesia is adjoined by the smaller Melanesia; to the east, by the gigantic island and water landscape of Polynesia.

At that time, most of this region had never been reached by a human being, the only exception being the Bismarck Sea, which is part of Melanesia and lies east of New Guinea: this area was already inhabited by hunter-gatherers, but these limited themselves to islands that offered sufficient hunting and foraging; the very smallest were thus ruled out. The Neolithics were subject to no such restrictions in their quest for land. While they, too, were unable to farm extensively on the small islands, they could at least do so to an extent that, together with fishing, allowed a reasonable standard of living.

The smaller the islands, the fewer generations it took to reach the point where no land was left for new families, and pressure increased on the youngest to go forth and discover new pastures. It goes without saying that these early seafarers boasted excellent nautical skills. But even the greatest seafaring talent was powerless against Pacific waves, which can be up to 20 metres high: in all probability, none of those who were claimed by them were among the ancestors of today's Oceanians.

Man swapping

For the hunter-gatherers at least, life on the Bismark Archipelago, which lay around 30 kilometres from the landmass of Sahul during the Ice Age and is now part of Papua New Guinea, was over. For tens of thousands of years they had gone about their business there undisturbed, until the newcomers suddenly appeared on the horizon aboard their sail-powered monsters, with their curiously tame animals in tow. The Austronesian farmers, whose ancestors had set out from Taiwan some forty to fifty generations before, had arrived. It was in this region that the oldest relics of the culture called 'Lapita' were found; they are about 3,600 years old.

The Lapita was the first agrarian culture in the South Pacific. In the space of 400 years, it had already spread via the Solomon Islands to the island chain of Vanuatu, and less than 100 years later to Tonga, another 2,000 kilometres further east. The Lapita occupied an area of around 4 million square kilometres – so the size of the European Union, only covered almost entirely by water.[1]

Opinions still differ as to how the people from Taiwan spread out into the South Pacific, and by what route their language and culture made its way across the Southeast Asian islands. One theory posits a fairly leisurely colonization from north to south: the immigrants settled on islands some of which were inhabited already by hunter-gatherers, then new settlers headed off to the next islands a few generations later. This assumption is based, among other things, on the genetic composition of the contemporary inhabitants of Vanuatu and surrounding islands, who carry both East Asian and New Guinean DNA – a strong indication of a slow expansion across the islands and interbreeding with hunter-gatherers.

In 2016 genetic analysis cast considerable doubt on this version of events. The DNA in question came from individuals who lived on Vanuatu 2,800 years ago and was obtained only thanks to a new extraction method. DNA had been found to be particularly well preserved in the hardest bone of the human body: the petrous bone in the temporal lobe of the skull. This discovery finally opened up large parts of the tropics to archaeogenetic research. The study of early farmers of the Lapita culture in Vanuatu and in Tonga, 2,000 kilometres to the east, yielded completely unexpected results: the DNA was entirely of East Asian origin, with no admixtures from Southeast Asian forager peoples. This in turn suggests that the Austronesian expansion didn't take place via the already inhabited large islands but away from them, without interaction with the Indigenous population, and was probably very rapid.

This theory of an initial expansion without significant mixing is also supported by another genetic finding. East Asian DNA didn't arrive on New Guinea until about 1,500 years ago, long after the beginning of the Austronesian expansion – and even then it didn't displace the hunter-gatherer genes. On the contrary, the farmers evidently failed to establish themselves in large parts of the world's second biggest island, as there are no traces of them in the genes of contemporary highlanders. All this points to some coexistence between foragers and farmers in this region; and this coexistence was probably characterized by trade – after all, the people of New Guinea had bananas and more to bring to the table.

Astonishingly, however, the inhabitants of Vanuatu today have an entirely different genetic signature from their ancestors of 2,800 years ago. They are no longer fully East Asian, but only 5 per cent – the large remainder of their DNA resembles that of the Indigenous inhabitants of the Bismarck Archipelago and New Guinea Highlands. This completely contradicts what genetic shifts throughout human history have consistently shown, namely that agrarian cultures always trumped those of foragers. Not so in Vanuatu: here, not only did East Asian DNA give way to New Guinean over time, but the archaeological trail of the Lapita culture, too, ended 2,600 years ago. So did the hunter-gatherers in fact displace the farmers? After appropriating their cultural techniques as a result of intercultural exchange? Against this speaks the fact that Austronesian languages are still dominant on the South Pacific Islands, 200 of them being spoken in Vanuatu alone. And it is generally the case with genetic displacements that the dominant incomers bring their language with them, because this is the key means of expression of any culture.

Presumably, then, the Indigenous genes were not completely displaced; there could have been instead a continuous gene flow between the Bismarck Archipelago and the smaller neighbouring islands of Vanuatu. This is because, if we assume that a constant and reasonably similar exchange between a very small island and a much larger one goes on over the centuries, or even the millennia, then it is only logical that the genetic structure on the small island will adapt to the one on the large island, where things remain in principle unchanged. A few migrants make no impact on a large population even over a longer period, but they do on a small one: it matters whether a drop of ink falls into a glass of water or into a bucketful. This effect is observable not only in the 95 per cent of New Guinean DNA carried by inhabitants of present-day Vanuatu but also on the islands further east.

It seems to have been mostly men from the Bismarck Archipelago and from New Guinea who married into the neighbourhood, at least judging by the dominance of their Y chromosomes. It is also perfectly possible that the same thing happened in reverse – that men from the smaller islands came to New Guinea to start a family; but no genetic evidence of such an exchange of Pacific manhood has been found. In the case of Tonga and Samoa, no such effects are observable in the genomes of present-day inhabitants, as these islands were probably too remote to allow a constant passage of people across the water.

The genetic data point to an enduring and reciprocal genetic bridge between the Bismarck Archipelago and Vanuatu that existed until thirty or forty generations ago. The gradual decline of the Lapita culture 2,600 years ago – in other words, the loss of the pottery skills that define the period archaeologically – may have been due to the small size of the island populations. After all, not everyone will have specialized in pottery at the time of the Austronesian expansion any more than we are all IT experts today. The number of those who practised and taught their traditional craft will have dwindled from generation to generation, as more and more people turned to an emerging technology vital for their survival: boat building. To put it bluntly, there was not much of a future in ornamental pots.

Experiments on the high seas

Today we have a pretty clear picture of the vessels people used to cross the Pacific during the Austronesian expansion, despite the fact that there are no archaeological finds from which to reconstruct their complex ships. Here – as indeed elsewhere – the discipline of experimental archaeology has played a key role. This specialist field first emerged in the late nineteenth century and seeks to answer, through experimentation, archaeological questions that are otherwise difficult or impossible to solve. In this case, for example, if a search of the ocean floor and of the Pacific islands fails to uncover any trace of a ship, researchers attempt to build one – using only the materials, techniques, and tools that would have been available at the time – and to sail in it to one of the islands. In this way, they hope to reconstruct the kind of vessel used by our forebears. Fortunately they are not starting entirely from scratch, given the existence of historical records of ships from which earlier models can be extrapolated.

Thanks to experimental archaeology – but also thanks to written records and oral traditions – we now have a rough idea of the impressive catamarans used by the early colonizers of the Pacific. Some of the most famous archaeological experiments were conducted by the Norwegian archaeologist and ethnologist Thor Heyerdahl, who died in 2002. In this way Heyderdahl brought the discipline into the mainstream. In 1947 he sailed from Lima to the Pacific on the Kon-Tiki, a raft made from balsa wood, in order to demonstrate his thesis that this part of the world could have been colonized from America. The voyage was successful and subsequently became the subject of a film, but it also showed the limited reliability of such experiments: the fact that they work today doesn't necessarily mean that the same methods were used in the past. But these experiments help us to understand the obstacles that faced the early conquerors of the Pacific – obstacles that experimental archaeologists contend with first hand.

Another impressive example of experimental archaeology is that of *Hōkūle'a* – a double-hulled catamaran in which the Polynesian Voyaging Society sailed in 1976 from Hawaii to Tahiti, 4,000 kilometres away, using nautical techniques from the time of the Polynesian expansion. Since then, *Hōkūle'a* has been used on multiple occasions, for voyages to places such as Micronesia, the Pacific coast of America, and Japan. The ship was quite openly built and used not just to conduct an experimental reconstruction of how the Pacific was colonized but also to revitalize the cultural awareness of our Hawaiian and Polynesian forefathers.

A fatal weakness for status symbols

On embarking, the Austronesian seafarers took with them the things they needed to survive. Along with dogs, pigs, and chickens, these included rats, which were presumably valued as a food source also because of their particular resilience on sea voyages. Coconut palms, too, came to the South Sea with the new arrivals. But farming conditions on the islands – especially the tiny islets of Micronesia – were of course nothing as good as in Eurasia – or even America, where they had fostered the rise of Neolithic cultures. Even today, there are few large areas of farmland to be found anywhere in the region of the Austronesian expansion, whose agriculture is largely based on growing root vegetables and rearing pigs and chickens – animals that don't need pastureland but are uncomplicated detritivores.

The islanders themselves had to adapt to this challenging environment. It may be, therefore, that each successive expansion to a remote island resulted in the expansion – in every sense of the word – of people whose metabolism was capable of storing large reserves of fat and hence of energy, to be used over the days or weeks of their long voyages. When it came to surviving long periods of famine during expeditions and getting through the initial groundwork necessary to establish an agricultural infrastructure, genes that made it possible offered a selective advantage. In those days they were a blessing for their carriers.

For their modern descendants, who are certainly not short of sugar, fat, or protein, they can become a curse by making them prone to obesity, with its associated health risks and lower average life expectancy. The fact that eight out of the ten states with the highest levels of obesity are located in the Pacific islands may be due to the genetic bottleneck caused by the Austronesian expansion – even if the relevant 'thrifty gene' has yet to be identified. Of the twenty countries with the largest proportion of obese people worldwide, eight are in the Arabian Peninsula, alongside the Pacific countries already mentioned. Again, most of these are regions where persistent long periods of drought and famine had to be overcome in the past millennia. The only country of the top twenty that bucks this geographical trend is the United States.

Although the Fiji Islands, Samoa, and Tonga yielded a relatively large amount of agriculturally useful land by Pacific standards, this land appears to have become insufficient around 1,200 years ago. It was then, if not before, that the last major stages of the Austronesian expansion began. Within 400 years, humans had colonized Tahiti, 2,400 kilometres to the east, and the Cook Islands, some 1,400 kilometres away. They advanced northwards to Hawaii and eastwards as far as Easter Island. The settlement of Polynesia, which covers 50 million square kilometres – more than North and South America put together – was little short of miraculous. In the case of Easter Island

– also known by its native name, Rapa Nui – they would have had to cross at least 2,000 kilometres of open sea in order to reach a piece of land just 24 kilometres wide.

It is not hard to see why the descendants of the Easter Island discoverers appear to have developed such a penchant for cult objects. After their incredible journey, their ancestors had good reason to call on supernatural powers and deities to guide and protect them. Over the centuries, the islanders carved extravagant stone statues from the volcanic rock; these were then distributed around the island – probably along specially made tracks – using ropes and tree trunks. It is thought that there were around 1,000 of these mo'ai; they weighed an average of 12 tons, and the largest remaining specimen is nearly 10 metres tall. The archaeological finds on Easter Island suggest that these artefacts served as status symbols for individual villages or tribes, each apparently attempting to outdo the other. On the slopes of the volcano Rano Raraku – the origin of all surviving mo'ai on the island – a spectacular unfinished specimen can still be found. It remains there to this day, as if waiting to be installed, like all its predecessors, but the 21-metre-long statue was never completed or erected.

Land of milk and honey

This abandoned giant positively symbolizes the fate of humanity: having reached Easter Island, the last bit of land left to discover, we came up at last against the limits that a finite planet imposes on a species that strives for infinity. Before human colonization, this place was most probably a densely wooded paradise. Yet by the time the Europeans landed here in the eighteenth century, all that was left was a barren steppe. Since the island was presumably settled less than 800 years before, it couldn't have taken more than a few generations of humans to turn virgin forest into a series of monocultures watched over by monstrous stone statues. There are various theories as to how this assault on natural resources came about, though all have the same basic premise: the islanders began to destroy the things they depended on from the moment they began to exploit them.

The principal victim of the settlers' axe was the honey palm, of which 16 million specimens are estimated to have existed before colonization.

This felling was done partly to clear land for agriculture and partly because the trees themselves were a valuable crop, yielding as they did a sweet and nutritious sap when sliced. There is even a theory that the island was deforested so excessively because the trunks were needed to transport the giant mo'ai statues. Either way, the formerly forest-covered ground would have been exposed to the tropical rain, which may have eroded and washed away fertile soil. It was probably an insidious but steady cultural decline accompanied by a dramatic fall in the population.

According to some estimates, only a tenth of the island's population at the height of its colonization was left by the time the Europeans arrived. The numbers living there during its cultural heyday in the sixteenth century were never reached again thereafter. The exploitation of the island by the Europeans and the infectious diseases they brought with them played a not insignificant part in this, and by the end of the nineteenth century there were barely a hundred Indigenous islanders left. Today the 8,000 inhabitants of the virtually treeless Rapa Nui depend mainly on tourism. Most of the goods they need are imported.

In earlier times, when Europe and large parts of Asia were paralysed by the horrors of the Black Death, trade flourished in the South Sea. The most intensive phase of this early transpacific economic area occurred in the thirteenth and fourteenth centuries, a period when all islands had been discovered and colonized but not yet fully exploited. There were trading routes between Hawaii and Tahiti, for example – two islands that lie 4,000 kilometres apart, with few useable staging points. Around 1400, however, the busy sea traffic came to an end and the islanders withdrew into themselves. In a world where even the smallest islands had been discovered and divvied up, leaving no room for new settlers, no incentive was left to embark on arduous, life-threatening voyages in search of a promised land beyond the horizon.

From then on, all efforts were devoted to squeezing the last resources out of the islands. This region, subjugated by Europeans in the seventeenth and eighteenth centuries and still a dream destination for global northerners with itchy feet, had long ceased to be an unspoilt paradise by 1400; in fact it was quite the opposite. In the Pacific, perhaps for the first time since leaving Africa, humankind had exploited its native planet to the hilt.

Fifteen canoes head for New Zealand

The final act of the Austronesian expansion was played out in New Zealand, which was reached by sail about 750 years ago, probably from the Cook Islands nearly 2,000 kilometres to the north. There, after the tiny islands of Polynesia, the conquerors encountered a much larger landmass. The migrants, who called themselves Māori, brought with them, among other things, the Pacific rat. Both species, humans and rats, were the first land mammals ever to set foot on the island: New Zealand had been surrounded by water for 40 million years, during which almost every ecological niche was occupied by birds, notably the flightless moa, which was wiped out in short order after the arrival of humans. New Zealand's eagle, 3 metres tall, also fed on moas, but soon enough it, too, became a victim of human settlement.

The natural environment of New Zealand – known as Aotearoa among the Māori – must have appeared positively welcoming to the newcomers: to this day, its flora and fauna are renowned for being the antithesis of Australia's hostile terrain. The birds, being totally unaccustomed to humans and hence not afraid of them, were easy to catch and kill, but there were no venomous or otherwise dangerous animals. In a place with no land mammals and with a favourable, humid climate, specializing in extreme defence mechanisms offered creatures no evolutionary advantage: such mechanisms, unlike in Australia, were simply not necessary.

According to the oral history of the Māori, whose ancestors came to the island only fifteen to twenty generations ago, settlement was accomplished with fifteen double-hulled canoes, each one representing a different Māori tribe. During the Musket Wars, which broke out in New Zealand in the first half of the nineteenth century and were facilitated by firearms imported by the Europeans, many of these tribes engaged in mutual hostilities, which caused the deaths of around 20,000 Māori. These conflicts were also about access to the island's natural resources.

Today there are some 700,000 Māori in New Zealand; they account for around 15 per cent of the island's population and are by far the largest ethnic group descended from Polynesian settlers in the Pacific. Asian migrants, who have been arriving in larger numbers since the 1990s, make up 10 per cent. Thus most of the genetic melting pot in this island country can be traced back to European migrants. A similar demographic shift also took place on Hawaii, where fewer than one in ten people are of Native Hawaiian descent: 40 per cent of the population in this US state have Asian roots, and 25 per cent have European roots.

Sweet potatoes with chicken

The Europeans' historical dominance in the Pacific is also a reason for some uncertainty as to whether the Austronesian expansion extended as far as the South American coast. While there is much evidence to support this view, one of the most obvious clues – the sweet potato – should nevertheless be treated with caution. This vegetable is common throughout the Pacific and unquestionably originated in the American Neolithic. This could point to a trade between the Indigenous inhabitants of the Pacific and the South or Central American west coast; on the other hand, it could have been popularized by the Spaniards who, when their fleet ruled the waves in the sixteenth and seventeenth centuries, operated a brisk trading route across the equatorial Pacific from present-day Mexico all the way to Taiwan.

Unfortunately few written records of this trade have survived, so we don't know whether the sweet potato first found its way to Oceania with the Spaniards or had already been traded with America a few hundred years before. A further clue to a Pacific–American connection at the time of the Austronesian expansion was furnished by some chicken bones found by archaeologists on the coast of Chile and dated to the thirteenth century. Chickens originally came from East Asia, so they wouldn't have been consumed in America at that time unless they were previously imported by sea. Here too, though, there is some uncertainty around the dating, so it's also possible that this was another souvenir from the Spanish.

The most reliable indication so far of an exchange between the Pacific region and the Americas comes from the Marquesas Islands north of Tahiti. The genomes of present-day islanders contain – albeit in minimal quantities – measurable genetic traces that have been interpreted as an early admixture from South America. Here at least, then, there may have been a connection, in the shape of a roundabout route via Easter Island. After all, any expedition starting from Rapa Nui in the uncharted east would almost inevitably have had to land in South America – even if that meant covering almost 4,000 kilometres. On the return voyage, the Humboldt Current that flows along the South American coast may have diverted some seafarers northwards, taking them from South America to the Polynesian islands close to the Equator. Perhaps they had women on board whom they had met in America – hence the genes detectable today on the Marquesas Islands. Also possible, of course, is that this route was taken by men who came as companions to female seafarers, or by people who migrated from America to the Pacific Islands out of a sense of adventure, sheer necessity, or itchy feet.

But this is only one possible explanation of how South American genes travelled more than 6,000 kilometres to the Marquesas Islands. As it is, the DNA traces suggest only a very sporadic exchange, which probably didn't even cover the whole of Polynesia: in the case of western Polynesia or Hawaii at least, there have been no such genetic finds in the Indigenous population. And there is nothing at all to support the hypothesis, often advanced in the past, that South America could have been partly colonized from the Pacific region. The ancient genomes from Tahiti discovered to date show no signs of an early genetic exchange with the Americas either.

To this day, very little is known about the route by which the Austronesian expansion arrived at its westernmost point, Madagascar. That it did so is well documented, however: for one thing, Malagasy, which is currently spoken by the majority of the almost 27-million-strong population, is the westernmost member of the Austronesian language family. And the DNA of contemporary islanders also testifies to this migration, which, according to the gene clock, must have taken place about 2,000 years ago. The average Madagascan genome is half African and half East Asian in origin. Both elements came at roughly the same time to the island, which is connected to Africa via the Comoros.

Sadly, there is as yet no ancient DNA from Madagascar, or any traces in the Indian Ocean that might give us a clue as to the Austronesian route. No Austronesian languages are spoken in the Seychelles, in the Maldives, or in Sri Lanka to the northeast, which would almost inevitably have been staging posts on any sea voyage to Madagascar. Besides, no archaeological remains have been found on the islands to date, let alone any DNA.

But the Austronesian seafarers may not have come to Madagascar by sea at all. Instead, the East Asian genes could have made their way there along the coast of India, Arabia, and Africa. This would account for the fact that African and East Asian DNA reached the island at the same time. Yet, if this version of Madagascar's settlement history were true, it could hardly have taken the form of a gradual expansion by land, since that would have taken much longer. The more plausible explanation is that the East Asian genes came via established trading routes to East Africa, and from there they washed up in Madagascar via the Comoros, as an Afro-Austronesian mix.

Adrift in the Caribbean

The conquest of maritime territory at that time was not confined to the Pacific and Indian Ocean: the same thing was happening off the coast of South America, too. In 2020 we saw the first successful attempt to decode ancient DNA from a large number of early inhabitants of the Caribbean, the genetic material being taken, once again, from the particularly stable petrous bone of the skull. The 300 or so samples came from several Caribbean islands of the Greater Antilles, notably Cuba, Hispaniola, and Puerto Rico, and were between 3,200 and 400 years old.

The first hunter-gatherers came to the Caribbean 7,000 years ago – by which route, we don't know. All we know is that each new leap must have seen them cover enormous distances – sometimes of around 150 kilometres. This may sound paltry in comparison with the distances involved in the Austronesian expansion, but this migration to the Caribbean happened much earlier: in those days there were no sails or catamarans, only simple canoes at best. Perhaps people even swam to the remote islands, clinging on to tree trunks and drifting for days and nights across the Caribbean Sea, in the hope of avoiding being dragged to the depths by currents or by predatory fish. We can be certain, at any rate, that they didn't leave the mainland just for the fun of it. The likelihood is that they ran out of fishing and foraging grounds on the South American coast – or they followed them from sandbank to sandbank.

The Caribbean petrous bone samples contained DNA both from the long-settled foragers and from the farmers, who arrived here around 2,800 years ago (Figure 32). They came from the Orinoco Delta in the

Figure 32 The first farmers in the Caribbean © Tom Björklund

The first farmers came to the Caribbean about 2,900 years ago. The region had been home to hunter-gatherers for many millennia, and they hung on to their territory long after the newcomers' arrival. There was almost no genetic mixing between the two populations.

south. Within a few hundred years, they had spread throughout the eastern Caribbean, to Puerto Rico and the Lesser Antilles. From there it's a relatively short hop to Hispaniola, the island now shared between Haiti and the Dominican Republic. But the farmers, who had previously caused the same kind of genetic upheaval on the Lesser Antilles as the Anatolian farmers had done in Europe, seem to have come up against an insuperable barrier on Hispaniola that kept them away for almost a thousand years. Archaeologists have long suspected that this barrier could have been a hunter-gatherer people that dominated the island and lived there in such numbers that the incoming farmers didn't stand a chance. This suspicion is now supported by the latest genetic data.

The Neolithics did ultimately manage to settle on Hispaniola, around 1,800 years ago. But, unlike before, on the Lesser Antilles, the foragers didn't give way to the newcomers. The two populations remained completely separate for a very long time: the genetic samples show hardly any mixing in the centuries that followed the Neolithic expansion to Hispaniola, and individuals with the typical signature of

Caribbean hunter-gatherer groups still existed in Hispaniola and Cuba into the eleventh century. This may have been due to a strong tendency to isolation in either group, or else to very small populations in each. Recent genetic analyses show that the people who lived in Hispaniola and the Bahamas 1,000 years ago, foragers and farmers alike, numbered scarcely more than 10,000. Although each population had strong family connections with the other island, there was no mixing between them.

Modern population estimates, which became possible only with the help of genetic analyses, are strikingly at odds with the millions of natives described by Christopher Columbus on arriving in Hispaniola in 1492; perhaps he hoped to secure support for further expeditions to the New World by grossly exaggerating the numbers to the Spanish court. One way or another, the number of Indigenous American inhabitants fell drastically after the arrival of Europeans, who brought numerous diseases with them, as well as enslaving or arbitrarily killing large numbers of Indigenous people.

This chronicle of subjugation, exploitation, and colonization is still inscribed in the DNA of present-day Caribbeans. Among Cubans, in particular the paternally inherited Y chromosomes are predominantly of European origin. Only the maternally inherited mitochondrial DNA can be traced – to the tune of 30 per cent or so – to the people who lived in those parts before the arrival of Europeans. This suggests that Indigenous men were largely prevented by the newcomers from fathering offspring.

The American–Russian axis

The quest for new territory not only drove humans across the entire Pacific to the Caribbean but also stirred things up in a region that can be life-threatening to the explorer in a very different way: the Arctic. As we know, our ancestors were the first people to fight their way to the cold north of Eurasia, the conquest of northern Siberia being a vital springboard for the exploration of America, which started 15,000 years ago.

North Alaska and the northern part of present-day Canada were first settled some 5,000 years ago, by the Palaeo-Eskimo (Figure 33). These have long been assumed to come from Central Siberia, 5,000 kilometres west of the Bering Strait. This theory is based on the existence

Figure 33 Advancing into the north of Alaska and Canada. © Tom Björklund.

Some 5,000 years ago, as people advanced into the north of Alaska and Canada, they developed new hunting and survival techniques. Some of these techniques are still used today by whale and seal hunters in the region.

of a linguistic bridge between the Na-Dené languages spoken by many Indigenous people in North America and the Yeniseian language family, whose only remaining branch is Ket. Nowadays Ket is spoken by scarcely more than 100 inhabitants of the central Yenisey Valley, which lies almost exactly in the middle of Russia.

The ancestors of this people probably specialized in hunting musk oxen and reindeer and spread to the Bering Strait about 5,000 years ago, at a time when the Neolithic was gaining ground in the south and the west of Eurasia. Although of course the strait had long since ceased to be dry, as in the days of the first settlement, it posed no great obstacle for the Palaeo-Eskimo. On the contrary, they made their home there, developing both its eastern and its western shores and ensuring a further busy exchange across the water. Also colonized at that time were the Aleutian Islands of southern Alaska. The ice-free portion of Alaska, which resembled the Russian tundra, served less as a habitat for the Palaeo-Eskimo at the time – at least there are hardly any fixed settlement structures from that period. More likely, the northern part of America was used as an extended hunting ground, to supplement the otherwise scant resources of the Arctic. The Palaeo-Eskimo had ventured as far as

Greenland about 4,000 years ago, but, again, rather on a seasonal basis: there are no permanent settlement structures here from that time either.

In archaeogenetic terms, the Palaeo-Eskimo are of huge significance. It is from them that we have the first ancient human genome ever to be sequenced; it was published in 2010, before that of the Neandertal. The sample was taken from 4,000-year-old hairs, and the results were ultra-sensitive. They demonstrated – or so it was believed at the time – that contemporary members of North America's Indigenous ethnic groups were not of Palaeo-Eskimo descent after all. The fact that the researchers who carried out the study were from Denmark, to which Greenland belongs as an autonomous country, made this even more explosive. On the basis of the genetic analyses presented in the study, it was postulated that, in reality, the so-called Neo-Eskimo – ancestors of today's Indigenous North Canadian groups, including the Yupik, Iñupiat, and modern Inuit – must have constituted a whole new wave of migration that had nothing to do with the Palaeo-Eskimo who were already living there. Given that the 'Danish' Vikings settled Greenland as early as AD 1000 from Iceland, while the 'Neo-Eskimo' didn't land there until the thirteenth or fourteenth century, this reading would make the Danes the ones with the oldest claims to the world's largest island.

Quite apart from the fact that it's rarely a good idea to base a nation's sovereignty claims on genetic data, the DNA analyses conducted at that time have since been partially discredited. While it's true that the ancestors of the Inuit and Yupik who are currently living in the Arctic arrived there no more than 800 years ago, they themselves were partially descended from the original Palaeo-Eskimo population. That much was demonstrated in 2019 by an analysis of the genome of prehistoric and modern inhabitants of Alaska, the Aleutian Islands, and Canada that was then compared with already known genetic data. The result was a whole new model of kinship relationships between Indigenous people that extend not just to the southern United States, but also to the Russian side of the Bering Strait.

According to this model, today's Indigenous inhabitants of the North American Arctic, as well as those of the Russian Chukotka Peninsula, can be traced back to the Palaeo-Eskimo. These forebears of the Inuit and Yupik appear to have crossed the Bering Strait three times in total. The first crossing took place when they came to Alaska as Palaeo-Eskimo

5,000 years ago and mixed with the people whom they found living there already. The second crossing occurred when they established the culture of Old Bering Sea on the Chukotka Peninsula and, over the course of 1,000 years, mixed with local groups on this side of the Bering Strait as well. The third passage followed about 1,200 years ago, again across the water and along the ice, towards Alaska and northern Canada. These people were genetically quite distinct from the 4,000-year-old Palaeo-Eskimo originally sequenced, but were nevertheless partially descended from that population. When they finally arrived in Greenland 800 years ago, they also brought with them the Thule culture associated with the Neo-Eskimo.

Around that time, the Vikings retreated from the island they had once christened Green Land, the Little Ice Age having turned it into an inhospitable place for them. Not so for the Neo-Eskimo, though: they were by then perfectly adapted to life amid the ice and were able to survive there even during that temporary phase of climate cooling. They developed above all a remarkable expertise in whaling, which is still cultivated by some of their descendants to this day. Armed with razor-sharp harpoons with blades made of stone or animal bones, these early Arctic inhabitants set off in their boats in pursuit of large and small whales. Once harpooned, the whales could no longer free themselves from the hunters' clutches: as soon as the exhausted animals were forced to come up for air, the water would once again turn blood-red. The drama ended with the hunters dragging their slaughtered prey ashore, where it would feed whole communities for days or weeks. The menu was supplemented with the spoils of seal hunts and, of course, daily fishing expeditions.

Foreign ships loom

With the conquest of the Arctic, the last chapter of our ancestors' expansion for the time being was completed. Eight hundred years ago, the oceans were part of human civilization, as was the perpetual ice in the north. In all probability humans would have reached Antarctica too, sooner or later; for it lies just 80 kilometres from Tierra del Fuego, the southern tip of the Americas, which had been inhabited by humans for 10,000 years. But this last great leap was not to be: time was not on their side. Five hundred years ago, the people of North and South America were overrun; monstrous ships pierced the horizon. Columbus landed in the Caribbean, Magellan circumnavigated Tierra del Fuego, and the Europeans went on to subjugate the Pacific as well. The tiny continent of Europe had just come through the dark, paralysing Middle Ages; now it was gearing up to conquer the world. The flowering of this small corner of the earth may have begun with the Neolithic, but would be almost unthinkable without the cultural influence of Asia. It was from there that humans began their march westwards, only to sweep across the entire globe as a weapon- and technology-wielding European power, armed with deadly infectious diseases and freed from all natural limits.

9

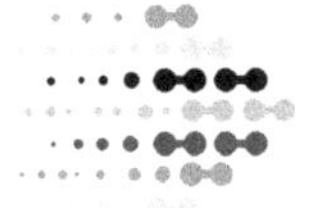

The Steppe Highway

During the Bronze Age, horsemen from the Asian steppes push both eastwards and westwards. China builds a fortress; Europe is overrun. The plague spreads, changing the course of history. Europeans suffer and spread the suffering.

Outsider odds

The people who colonized the entire Pacific and parts of the Indian Ocean during the Austronesian expansion may have been the most talented and daring seafarers our species has ever produced. But their trajectory ended on the islands, which – fishing expeditions aside – became their home and never gave rise to seafaring nations. Nor could they, given the lack of natural resources for shipbuilding. It was the Spanish and the Portuguese who rose in the fifteenth and sixteenth centuries to become rulers of the waves. In the space of a century, they conquered Central and South America, changing the genetic map there dramatically. In the early seventeenth century the British took over as the leading sea power and spent the next centuries building an empire that spanned almost the entire globe. Thereafter, the northern part of America was dominated by Western European DNA: there too, the Indigenous inhabitants were marginalized, which often meant annihilation. In addition, the slave trade that was to follow brought African ancestry to the New World – particularly to the north, but also to Central and South America. This is how America became the continent with the greatest genetic diversity after Africa.

And yet it was anything but a foregone conclusion that Europe would be a producer of world powers. The smallest continent after Australia, it

Figure 34 The Steppe Highway.

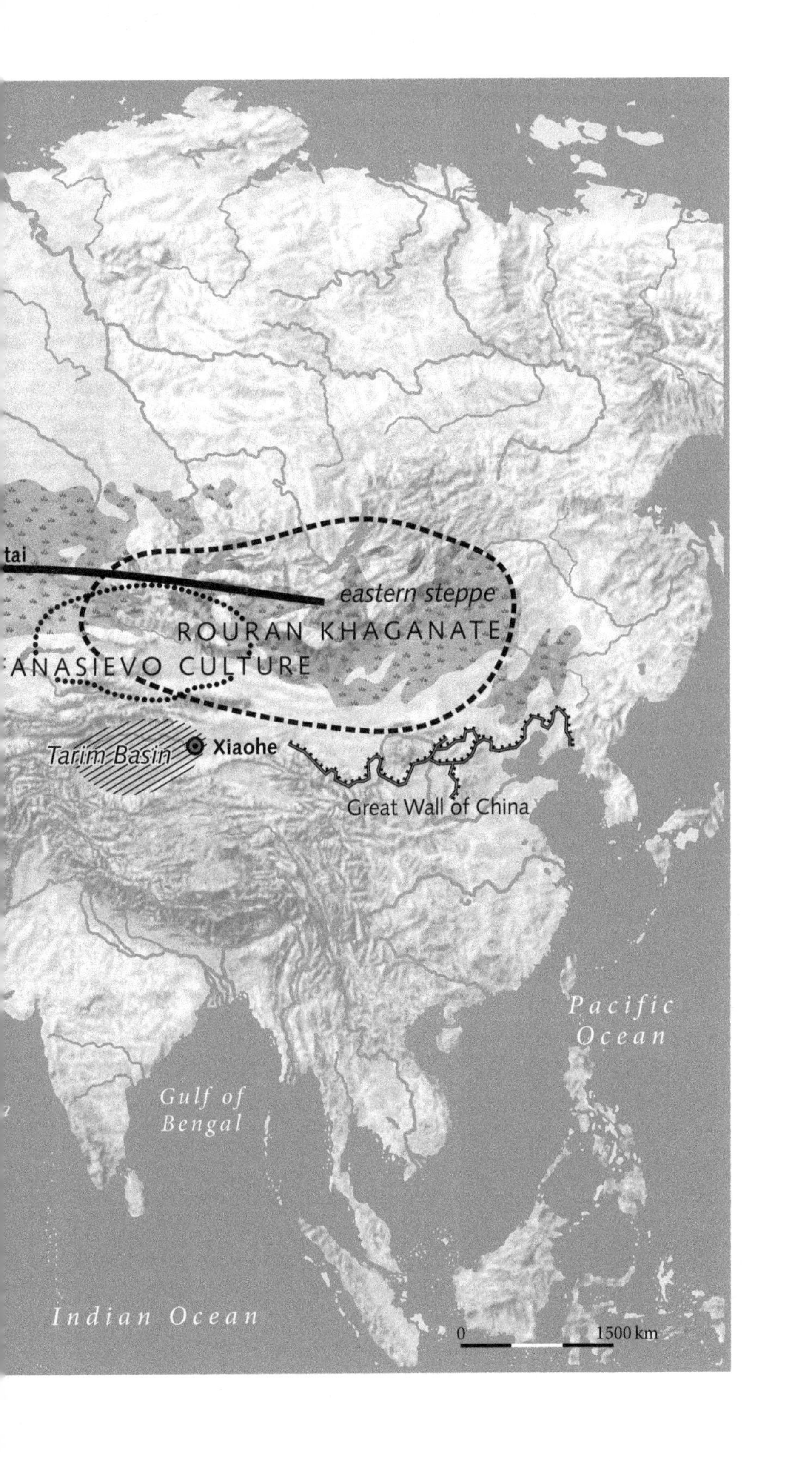
tai
eastern steppe
ROURAN KHAGANATE
ANASIEVO CULTURE
Tarim Basin
Xiaohe
Great Wall of China
Pacific
Ocean
Gulf of
Bengal
Indian Ocean
0
1500 km

lies at the periphery of the Eurasian landmass, and most migrations in early human history initially gave western Eurasia a miss. From Africa, migrants generally headed for East Asia – particularly given the enduring dominance of the Neandertals in Europe. Climatically, too, Europe was a challenging place, although this disadvantage was offset by its rich hunting grounds. The rise of great Neolithic cultures and the resulting series of emerging agrarian powers were played out exclusively – and with good reason – at the climatically favoured latitudes that stretch from North Africa through the Fertile Crescent and the Indian subcontinent to Central America.

In Asia and Europe alike, the Neolithic benefited from a seemingly custom-made fauna, which could no longer exist in Africa thanks to the coevolution of its wildlife with human predators. The virtually endless landmass of Eurasia allowed the spread of domesticated crops along climatically similar latitudes and provided a constant supply of agricultural land and room for expansion for thousands of years. As a result of the population growth that began with the Neolithic and was subject to far fewer natural constraints from then on, this surplus of resources was accompanied by a surplus of labour. The division of labour that ensued allowed for the development of increasingly specialized crafts as well as military arts.

The latter in particular experienced a boom with the dawn of the Bronze Age, when the practice of alloying tin and copper first began, some 5,300 years ago, in the Middle East; and it was there, too, at the southwesternmost tip of Asia and on Europe's doorstep, that the Iron Age was born, more than 3,000 years ago. This steady progress relied on a flourishing trade, as the precious raw materials occurred only in a few favoured parts of the continent, and never all in one place. Alongside goods, know-how was also exchanged. With this transfer came a genetic homogenization of Eurasia's populations, which had been largely segregated at the beginning of the Neolithic, and above all a technological race between competing and cooperating systems, which otherwise risked being dominated by neighbouring empires.

But none of this is enough to explain Europe's subsequent rise to the top of the global power structure. Without a doubt, the subcontinent offers ideal conditions on account of its geography: surrounded by water, it offers neighbours access from all directions. But there was water elsewhere too, and the Austronesian expansion clearly showed that East Asia would make an excellent point of departure for seafaring nations. Even in the early fifteenth century, long before the Europeans, the Chinese Empire already commanded a large merchant fleet. Its eight-masted ships may not have been 120 metres long, as claimed in historical accounts, but they were certainly significantly larger than the ships later used by the Spanish, the Portuguese, the Dutch, and the British. The legendary fleet of Admiral Zheng He transported not just goods but also soldiers, thereby reinforcing China's claims to power; some expeditions were more than 300 ships strong. The Chinese trading zone extended from the western Pacific to the East African coast, which could be reached with great accuracy thanks to China's invention of the compass back in the eleventh century. Legend has it that the inhabitants of East Africa were less than impressed when, at the end of the eleventh century, a plethora of much smaller ships appeared off the coast; their sails must have looked rather puny in comparison with the familiar Chinese ones. In short, the sea route to India discovered by the Portuguese explorer Vasco da Gama was not the first trading route to the subcontinent.

And yet, despite all this, it wasn't the Chinese who went on to discover the New World on the other side of the ocean. Around the middle of the fifteenth century the project of naval supremacy lost the support of China's rulers. According to one popular interpretation of events, this was due to a desire to prevent the rise of a rival merchant elite. Other theories argue that the ships were costing the empire too much and hadn't brought in anything like the amount spent on them. Whatever China's reasons were for withdrawing from international maritime trade and for refraining from any further investment in a high-powered fleet, the opposite was happening at the other end of Eurasia.

By the time Columbus landed in America, da Gama circumnavigated the Cape of Good Hope, and Magellan crossed the Pacific, Chinese supremacy in the Indian Ocean had all but disappeared. Barely two generations had elapsed since the Chinese Empire had abandoned its seafaring ambitions. No one can say how world history would have developed if things had taken a different course in East Asia at that time – and if the Chinese had sailed around Africa and gone on to land in Europe, or done battle at sea with the Portuguese galleys. Which just goes to show how much the trajectory of human history depends not just on accidents of fate, but also occasionally on human caprice. Europe was, at all events, no obvious candidate for a global ascendancy that ended up transforming human genetics from America to Australia. But, once the decision was made, it was irreversible.

The beautiful mummy and the cheese

The genetic history of China stands out because, although this empire in the east had traded with the whole of Eurasia since antiquity, via the Silk Road, the Chinese tended to keep themselves to themselves, ensuring genetic continuity from the first Neolithics of East Asia to contemporary Han Chinese. This isolation was favoured by the natural barriers of the Himalayas and of the Gobi Desert; and from the seventh century onwards these barriers were supplemented by fortifications that can be regarded as forerunners to the Great Wall of China erected in the Middle Ages. Built over the centuries, those structures were designed to keep away the hordes of mounted warriors from the northern steppe, an attempt that met with mixed success. At any rate, the invaders left few genetic traces, if today's Han Chinese are anything to go by. Things were different in the other direction: subsequently East Asian DNA spread not just along the route of the Austronesian expansion but also into western Eurasia to some extent.

Indeed, the western part of present-day China – the autonomous region of Xinjiang – came under the control of the empire only in the first century BC and changed hands repeatedly thereafter. This is reflected accordingly in the genes of its present inhabitants: although they are more closely related to west Eurasians, the Chinese component is clearly present and reveals an east–west divide. The DNA map of Xinjiang testifies to migrations from Central Asia that link a large part of the population genetically to western Eurasia. The political implications of this finding are clear: Xinjiang is the autonomous region of the predominantly Muslim Uyghurs, who in recent years have come under considerable and often violently enforced pressure from the Chinese government to assimilate. This is despite the fact that Han Chinese represent 40 per cent of the region's population and the Turkic-speaking Uyghurs around 46 per cent.

In ancient times, the western migrants' gateway to Xinjiang was the Tarim Basin. This basin is surrounded by high mountains on all sides except the east and is roughly 1,200 kilometres long. About 4,000 years ago, this barren, partially desert-like region was a veritable lakeland, fed continuously since the beginning of the Holocene by the melting glaciers of the Himalayas. In the basin, the water had nowhere to go, unlike in the south, where it flowed via the Ganges into the Indian Ocean. As a result, humans found much to attract them here from the early Holocene onwards, and duly settled along the shores of the many lakes.

Evidence of this development comes in the shape of the world's oldest 'dry mummies', which were discovered in 2004 in the eastern part of the Tarim Basin. These mummies were buried about 5,000 years ago and were perfectly preserved in the dry climate. So perfectly, in fact, that one notable burial object remains intact to this day: a piece of cheese. Protein analyses of the find confirm that this object, which was hung around the neck of the 'Beauty of Xiaohe', is indeed a fermented milk product. Dairy production was, it seems, a fundamental part of the lives of early settlers in the Tarim Basin – unlike grain cultivation, of which no traces have been found. Alongside dairying, fishing, too, was big in this region.

In 2021 scientists sequenced DNA from the Tarim mummies. The results showed on the one hand a genetic continuity with the hunter-gatherers who had lived there previously. This suggests a regionally autonomous trend towards pastoralism that appears to have been influenced only minimally by Iranian or Chinese Neolithics. On the other hand, the analyses revealed a genetic discontinuity with present-day inhabitants of the region, who do carry elements of the DNA found in the mummies, but only to a very small extent.

The signature of the cheese-loving Tarim Basiners was overwritten by the steppe peoples that arrived later from the west and for which the high mountains were evidently less of a deterrent. The people of the so-called Afanasievo culture were the early protagonists of a millennia-long development that shaped the history – genetic and otherwise – of almost the entire Eurasian continent.

In the cold steppe landscape, which proved to be largely inaccessible to the Neolithic revolution, powerful nomadic cultures built on animal husbandry evolved just over 5,000 years ago. Their greatest asset, however, was an animal that resisted domestication everywhere else: the horse – a loyal human companion and one of the most valuable military aids up until and during the First World War. The barren central part of Eurasia was to become home to an endless succession of equestrian peoples who found their way to Europe along the 'steppe highway', from Mongolia in the east to the Hungarian *puszta* in the west. For anyone with a tame horse at their disposal, this seemingly infinite grassland offered boundless riding with hardly an obstacle in the way, the Carpathian and Ural Mountains being among the few exceptions.

On horseback and underground

By the time the horsemen of the steppe cultures reached western China, they had already covered a great distance – and at a hitherto unimaginable speed. Unlike other migrants before them, the steppe peoples did not, it seems, disperse across the steppe gradually, from generation to generation. They appear instead to have used the highway purely in order to establish a nomadic culture a thousand kilometres to the east. The latest analyses support the long-standing theory of a close affinity between the Afanasievo and the Yamnaya cultures: both existed simultaneously in the Pontic steppe north of the Black Sea, and genetic data reveal the two to be very closely related. In other words, the people who shaped the Afanasievo culture between the Kazakh steppe and the Altai Mountains had their origins in the west. This means that there were only two to three generations between their departure from the Black Sea and their arrival in the Altai Mountains, 3,000 kilometres to the east. Such an unprecedented rate of migration could have been accomplished only on horseback.

The domestication of the wild horse is thought to have been first achieved by the Botai people, a culture that emerged 5,700 years ago. The name derives from an archaeological site in the north of present-day Kazakhstan. In the 1980s, pit houses were excavated there whose design, though familiar from other prehistoric settlements, was particularly common in the steppe. The settlements of the Botai culture were predominantly underground, mainly as a result of the conspicuous lack of wood in the steppe – a material without which it is hard to build vertically. Since its discovery, the Botai site has been intensively investigated and hundreds of thousands of bones have been found in and around the hundred or so settlements. Nearly all of these are from horses, on which the Botai appear to have been positively dependent. These animals were used for more than herding sheep: in Kazakhstan mare's milk was (and is) regarded as a delicacy, partly because of the speed of the fermentation process in warm conditions. Known as *kumis*, this drink tastes very much like kefir but also slightly like beer, because it has 1–3 per cent alcohol content.

The settlements of the Botai culture were often dozens of kilometres apart, so the pastoralists and their herds certainly didn't get in each other's way. Anyone wanting to exploit such vast areas for dairy farming and to keep control over them was dependent on the horse. Over time, people learnt how to be one with their animals. Riding lessons were presumably part of one's early education, and by the time they reached their teens the children of the steppe would have been able to ride hands-free – a skill that was to become a killer application for later nomadic equestrians. To be able to strike your enemy down, with bow and arrow, while being carried by the world's fastest vehicle was to possess the most terrifying weapon of war imaginable.

The Botai never got to reap the benefits of this innovative edge, though: their trail disappeared more than 5,000 years ago, before the great expansions east and west along the steppe highway. The art of horsemanship was probably perfected a little later, by the people of the Yamnaya culture, which arose in Eastern Europe, in the Pontic steppe north of the Black Sea, and lasted much longer than the Botai. The Yamnaya people cultivated a pastoral lifestyle as well. In their community the horse was valued far above other domestic animals, being used not only for herding but, importantly, to venture into areas that were inaccessible on foot.

Invasions

The takeover of western China by the Afanasievo culture was only one of the eastern offshoots of the incipient steppe expansions. Around the same time the pastoralists were also advancing into Europe. It was a wave that was to change the genetic structure of that subcontinent forever: for an equivalent upheaval to happen today, 10 billion people with foreign DNA would have to come to Europe, or 1 billion to Germany. From an area between the Pontic steppe and present-day Ukraine, the migrants journeyed west, which led in a few centuries to a wholesale genetic shift, not unlike the one associated with the beginning of agriculture in Europe. The DNA of the people who, starting 4,900 years ago, brought the Corded Ware culture from Belarus to the Rhine and the DNA of those who, less than 400 years later, imprinted the Bell Beaker culture all the way from the British Isles through Central Europe to the Iberian Peninsula can be traced back in large part to this wave.

Naturally, the people who came to Europe from the steppe and settled there adapted their lifestyle to their new environment (see Figure 35). Within a few generations, the nomadic horsemen, who also brought with them the craft of bronze working from eastern Europe, had become farmers. They often made use of the existing infrastructure. The most famous example is Stonehenge in England, which was seamlessly adopted by the incomers, most likely as a place of worship. They also appear to have emancipated themselves from the horses domesticated by their ancestors in Central Asia, as these were evidently returned to the wild. It is from these liberated animals that modern Przewalski's horses are descended: the latter are not European wild horses, as has long been assumed, but are genetically traceable to the animals domesticated in the steppe by the Botai culture. In Europe the steppe migrants didn't abandon riding, however: they probably just imported horses from Eastern Europe instead. Then, some 4,000 years ago, the Sintashta culture domesticated the animals that eventually gave rise to our modern domestic horse. Archaeogeneticists showed in 2021 that all modern domestic horses descend from these Sintashta horses. Within a few hundred years they spread all over western and central Eurasia and changed our methods of transportation and warfare forever.

The new sedentism of the incoming nomads did nothing to diminish the importance of the steppe highway – it's just that now it was no longer a one-way street. With the Bronze Age, this corridor began – not least through increasing trade – to funnel genetic components from Central Europe back to the east. Via Central Asia, these genes eventually reached the Altai Mountains, which seem to have presented an impossible barrier to all the movements of expansion to the east during the following millennia.

In time, this reverse trend from west to east brought people from Europe to Central Asia – people whose representatives, at least 4,100

Figure 35 Reconstruction of a settlement belonging to the Sintashta culture
© Rüdiger Krause

Reconstruction of a settlement from about 4,000 years ago that belonged to the Sintashta culture and was situated at the foot of the Ural Mountains (Arkaim site). Several dozen 'terraced houses' arranged in an inner and outer ring were protected by a 5-metre earthen wall around the settlement, which accommodated up to 2,000 people.

years ago, forged the Sintashta culture, which spread across present-day southern Russia to Kazakhstan. The people of Sintashta culture worked and traded with copper, and this resulted in close economic ties with the rich countries of the Middle East, where metal was in great demand. They were also the first to build larger settlements and fortifications in the steppe. Their culture gave rise later to the Andronovo culture, which stretched at times from the Caspian Sea to present-day Mongolia – until 3,000 years ago. Unlike its predecessor, the Andronovo culture has left no traces of larger settlements: its own consisted rather of a few post-built houses whose supporting elements were driven one metre into the ground and then filled in above ground level with wattle and daub.

Impressive architecture – of the kind already found aplenty in other parts of the world – this was not. On the contrary, these settlements were little more than spartan crash pads for herders used to spending most of their days on horseback.

Nothing to inherit, everything to gain

The huge expansion of the Andronovo culture is further evidence of how little incentive there was to build permanent settlements in the steppe and of how keen anyone looking to start a new life would have been to do so elsewhere, preferably within rideable distance from their relatives and possibly together with their nomadic horsemen friends. Understandably, the availability of abundant pastureland made the rigours of sedentary farming appear somewhat less attractive. This is not to say that some didn't give it a try.

Notable among those who did were the people of the Srubna culture, which was spread across the Pontic steppe and is closely related to the neighbouring Andronovo culture. In that region, agricultural structures were established some 800 years after the end of the Yamnaya period. The settlements of the Srubna people typically consisted of semi-subterranean pit-houses and were also used to store grain. That period ended about 3,200 years ago, however, and was to be the last serious attempt to settle in the steppe in the Neolithic tradition. It was an experiment that, as we will see, may have had fatal consequences for the subsequent course of European and Asian history.

Like the Srubna and the Sintashta, many other cultures of the steppe were short-lived, in some cases only lasting a few generations. In Africa and the Middle East, on the other hand, stable empires had long since arisen from the explosion of wealth created by the Neolithic revolution – notably those of Egypt and Mesopotamia. These were a much greater source of wonder to archaeologists than the modest remains of steppe dugouts. Which is not to say that the steppe peoples weren't responsible for any significant cultural achievements: there was bronze working, the domestication of the horse, and the development of the chariot for a start, not to mention the impressive burial mounds typical of so many steppe cultures. Bronze work and kurgans (burial mounds) then duly became features of the European Corded Ware and Bell Beaker

The dominance of horsemen

The migration from eastern Europe appears to have been, like so many before it, a largely male-dominated venture, with eight out of ten migrants likely to have been men. That much can be inferred at least from the distribution in Europe of X chromosomes (i.e. one of the two types of sex chromosome, X and Y; women have two X chromosomes, men one). The average X chromosome shows significantly less steppe ancestry than the other chromosomes, meaning that it must have been passed on much more frequently by Indigenous farming women. And we find the same pattern of a male-driven migration reflected in the Y chromosomes of modern Europeans too (Y being the chromosome that determines maleness). In some parts of the continent, for example the United Kingdom and Ireland, more than 80 per cent of Y chromosomes derive from the steppe. By contrast, no such shift occurred in the case of the maternally inherited mitochondrial DNA.

All this evidence points in the same direction: the children born to European women in the early Bronze Age must have been conceived almost exclusively with the newly arrived migrants. What became of the Indigenous Y chromosomes, we don't know. There is no doubt, however, that the armed horsemen would have won hands down in any conflict with the native farmers. For us today, the tiny and otherwise relatively insignificant genetic locus of the Y chromosome provides an unmistakable clue as to the outcome of this competitive situation. In the rest of the genome of the average modern European, the steppe component is not quite as dominant: in eastern and northern Europe, where the migrants arrived first, it accounts for roughly a half, decreasing in favour of the Neolithic component the further south you go. This is particularly true of the Iberian Peninsula, which was the last place to be reached by carriers of steppe DNA. The 'original' DNA of the European hunter-gatherers represents the third pillar in all parts of Europe, but generally accounts for the smallest proportion.

cultures – and thereafter of the Central European Únětice culture, to which they gave rise and which bequeathed to us the world-famous, approximately 4,000-year-old Nebra sky disc found in Saxony-Anhalt, Germany.

Given the barren conditions of the steppe, one cannot but be surprised at the rapid prosperity of its people over the course of millennia, a phenomenon that subsequently led, among other things, to the vast empire of Genghis Khan. All this was plainly not based on a flourishing agriculture or a wealth of mineral resources. No: if you wanted to get anywhere in the steppe, you had to look beyond the horizon – not just to keep tabs on the cattle grazing there, but to look for a new life. Consequently the outlook of a steppe dweller would have differed fundamentally from that of your average European or farmer in the Nile Delta, but not so much from that of the people who set out 200 years ago to cross the vast expanse of America with horse and cart, to reach the Pacific. Steppe dwellers grew up in the most unpropitious environment imaginable, and had almost no choice but to become pioneers in the course of their lives.

Relying on one's parents' inheritance was rarely an option. For one thing, there was no agriculture to speak of in the steppe, nor was there any shortage of land. Unlike in the once densely populated hotspots of the Neolithic period, even the most powerful steppe ruler could hardly have founded a dynasty based on the accumulation of land, as this resource was unlimited – and not particularly valuable at that. True, copper and bronze were precious commodities in the steppe as elsewhere, and the impressive burial mounds of certain individuals leave no doubt that extreme riches were amassed here too – perhaps even more than in other parts of the world. But these were material, tangible goods, built up over a lifetime, not princedoms; nor did they typically go to one's offspring but rather into a swanky tumulus.

In Europe, meanwhile, structures had already been established that may be interpreted as early signs of a patriarchal social system. We know this from genetic analyses of more than a hundred individuals from the beginning of the Bronze Age who were buried in multiple settlements along the river Lech in southern Germany. According to the genetic data obtained from the graves, a large majority of the women buried there came from outside the settlements, as opposed to only one of the men.

If this region is representative of life in Bronze Age Europe at that time – as there are good grounds to believe – it seems that women in those days left their home as teenagers in order to marry elsewhere. Men, by contrast, appear to have taken over their fathers' farms whenever they could: the graves were found to contain dozens of sons from identical households, but no grown-up daughters (see Figure 36). Also striking was the fact that burial objects were far more generously bestowed on native women and on women married into the family than on outsiders with no family connections: those were probably workers, or perhaps serfs. All these factors are suggestive of settlement and family structures that were grouped around patriarchs.

As elsewhere, in the steppe too, children whose parents were able to pass on the necessary tools – be it powerful horses, a large herd of

Figure 36 The beginnings of a patriarchal society © Tom Björklund

In Central Europe, the beginnings of a patriarchal society are observable in genetic data that go back to the early Bronze Age. There are strong indications that women began at that time to leave their families in their teenage years to marry into other households.

cattle, or expert craftsmanship or horsemanship – undoubtedly had a better start in life. More than elsewhere, though, it was ultimately up to individuals to take control of their destiny and carve out a career for themselves. Courage alone wasn't enough. Unlike the Austronesian conquerors, for example, who were busy discovering uninhabited islands in the Pacific at the time, the horsemen of the steppes couldn't rely on finding virgin territory, whether they headed east, west, or south. Anyone riding off alone here into the sunset had to be prepared for their arrival on horseback to cause wonder, but perhaps at the cost of their lives.

Ensuring numerical superiority was a must in such expeditions, which sprouted from the steppe in all directions in the course of the coming millennia. Charismatic leaders, who perhaps journeyed from one settlement to the next in search of young, land-hungry associates, were the ones with the best chances. Steppe dwellers may have lived far from civilization but, thanks to a continuous exchange with the rest of Eurasia from the Bronze Age onwards, they were nevertheless fully aware of the deadly power wielded by their combination of bronze weapons, chariots, super-swift horse bows and horses, as well as of the riches they could capture with them. Whatever promises a tub-thumping chieftain might make, everyone knew that, with luck on your side, there were even better rewards to be had.

Indian elites

The steppe migration that turned Europe upside down 4,900 years ago rested on a numerical superiority undeniably demonstrated by the genomes of modern humans. This was evidently accompanied by a clear cultural dominance, which manifested itself not just in the establishment of the Corded Ware and Bell Beaker cultures but also in European languages. It was from the steppe that the Indo-European family of languages originated, displacing virtually all those spoken hitherto; the only one to survive was Basque, whose forerunner is thought to have come to the Iberian Peninsula with the Neolithic revolution, more than 7,000 years ago. Language displacement is one of the surest signs of cultural hegemony, which is nearly always expressed in the spoken and, in due course, the written word. Were it not so, there would be abundant historical examples of conquerors of foreign lands adopting the local tongue.

This phenomenon is particularly striking in the case of India. According to the latest genetic data, the steppe expansion reached the Indian subcontinent about 3,600 years ago, at the height of the Andronovo culture. While Hindi – derived from the Indo-European language Sanskrit – is dominant in northern India today, the proportion of non-Indo-European Dravidian languages is gradually increasing towards the south of this 1.3-billion-strong state. This trend is mirrored by the distribution of steppe DNA, which in the north accounts for an average of up to 30 per cent, while in the south it is less than 5 per cent.

However, this variation in the steppe component is observable not only between northern and southern India but between certain sectors of the population too. And, ironically, in the north and in the south alike, a higher-than-average steppe component is consistently found in a socially privileged group: the highest 'caste' of the Brahmins, who traditionally make up a large majority of the Hindu priesthood. Whereas in other castes the proportion of steppe DNA declines from north to south in line with the population average, the genes that came in from the north are common to Brahmins throughout the subcontinent.

This genetic finding is politically not unproblematic, to say the least. Even though the caste system no longer operates officially as a social classification system in India, it continues to pervade everyday life, from birth through marriage to death. As the highest caste, the Brahmins have a special role to play in the Hindu tradition, being teachers and scholars of the Vedas, a collection of religious texts that were passed down and subsequently transcribed. These texts also record a migration to India from the north, specifically about 4,000 to 3,500 years ago. Those migrants are sometimes referred to in the texts as 'Indo-Aryans' – a clear indication of the elevated status of the Brahmins, who also stand out from many other Indians through the lighter colour of their skin.

Even though the new genetic findings do not point to a coherent group of Brahmins, but merely to an above-average dominance of the migrant component, they could nevertheless encourage some traditionalists to use molecular data from the latest biotechnological research to defend the outdated caste system. Even so, the fact remains that most Brahmins also have their genetic roots in south Asia.

A fateful mutation

Unlike in India's case, for Europe there are no oral, let alone written, records of the era of the great steppe migration. Moreover, there are no archaeological finds, and hence no ancient DNA from the early steppe migrants – not least because most of Europe appears to have been a no man's land 4,900 years ago. At any rate, there are scarcely any skeletal remains from this time for an interval of 150 years, and no remains of any settlements either: when the eastern migrants rode into Central Europe, they found themselves in a remarkably unpopulated region. What became of the few people they encountered, we don't know, but there are no signs of any major battles having taken place.

We do have clear evidence of something else, though – something that may have had a deadlier impact than the sharpest bronze sword or the swiftest horse: the Stone Age plague. The earliest known plague bacterium was sequenced in 2017, from bones that were buried 4,900 years ago in the Pontic steppe – the same region from which the steppe migration is thought to have originated. Indeed, new genetic analyses of Stone Age bacterial strains from all over Central and Eastern Europe have since consistently confirmed the suspicion that the Stone Age plague spread via the same route as the eastern migrants shortly before, and subsequently found its way back again 3,600 years ago, when people began to return, eventually reaching the Altai Mountains. All this suggests that the plague smoothed the migrants' passage. The sudden disappearance of European settlements during the 150-year dark period could have been caused by the genetically verifiable, sporadic contacts that existed before, such as those between people in the Pontic steppe and in the region of present-day Bulgaria.

In the Stone Age plague, however, as genetic analyses have shown, the genome of the bacterium lacked a key component that was later responsible for the deadly impact of the Justinianic plague in the sixth century and the Black Death in the fourteenth. The Stone Age pathogen is thought to have caused a pneumonic plague that spreads via the respiratory system when an infected person coughs up tiny particles from their lungs. By contrast, the bubonic plague that came later had a much more efficient transmission route: via the then omnipresent fleas that passed between rodents and humans. Mutations in a number of virulence genes of the bubonic plague bacterium caused the guts of fleas that had ingested infected blood to become blocked by coagulated blood clots. From then on, every drop of blood that the flea ingests from a new host is contaminated with plague bacteria and instantly regurgitated into the host. This process is repeated for as long as the flea, driven mad with hunger, keeps infecting new victims – until eventually it starves.

In these circumstances, there was a good chance of an outbreak of bubonic plague wherever large numbers of humans and rodents coexisted – and, hygiene being what it was in those days, there was no shortage of fleas. The earliest known bubonic plague genome was published in 2018: it was 3,800 years old and originated in the region around the southern Russian city of Samara – in other words, from a time and place that saw the first attempts at grain cultivation in the Sintashta and Srubna cultures. The harvested grains also had to be stored, which at that time would have led almost inevitably to rodent infestations. It is conceivable, then, that the foundations for the mutation of the pneumonic to the bubonic plague bacterium were laid here. While bubonic plague is slightly less deadly than pneumonic plague, it is much more contagious. And the rodent populations of Central Asia remain one of its most persistent reservoirs to this day.

Soon after the emergence of bubonic plague, the cultures of the steppe collapsed – notably that of the Srubna people. There is much to suggest that the bubonic plague bacterium also found its way into other parts of the European continent over the next few centuries via the steppe highway, and it's probably only a matter of time before further evidence of this is found in ancient bone samples. There are clear signs of a westward spread in particular, not least to the power centres of the Middle East at the time.

At the time of the first outbreak of bubonic plague in Central Asia, those centres were the Assyrian Empire between the Euphrates and the Tigris, the New Kingdom of Egypt that extended as far as Lebanon, the Hittite Empire in Anatolia, and the Mycenaean and Minoan civilizations in the Aegean. But 3,300 years ago this region was thrown into turmoil, if not panic, at least according to many accounts of that period. These mention repeatedly enemies who came from over the sea, who are referred today in archaeology as 'the Sea Peoples'. The Hittite Empire fell around 1,200 BC and the neighbouring empires followed like dominos over the following decades: texts of the time allude to the 'Hittite plague'. As in Europe at the time of the steppe migration, here too things fell apart, leaving scarcely any historical records or documents for around a century and a half – just ruined cities and states.

In 2019, DNA analyses from Lebanon and Israel backed up contemporary accounts of the invading Sea Peoples, reinforcing the suspicion that this episode could have been connected to a plague pandemic. After the arrival of the Sea Peoples, a distinct change was reported in the local population. People who lived there in its wake exhibited a new genetic component that had not existed before in the Levant and that came from southern Europe. Perhaps it was the plague here too, then, that decimated the Indigenous population and paved the way for a new settlement.

If in the East you don't succeed...

After the bubonic plague, not much else happened in the steppe for many centuries. Then, about 2,800 years ago, the Scythians appeared on the horizon – an equestrian people with a nomadic lifestyle that abstained altogether from permanent settlements and no longer practised even simple forms of agriculture. Even more than previous steppe cultures, theirs centred entirely on horses, which they used to build an empire that extended at times from eastern Europe to the Altai Mountains. The Scythians – themselves a multiethnic state – were genetically very different from the peoples that had inhabited the steppe a few centuries before. This, too, suggests a preceding population decline due to the plague. And the Scythians were only one in a whole series of equestrian tribes that repeatedly invaded Europe until well into the fifteenth century. The fact that these horsemen set their sights chiefly on the west had to do on the one hand with the wealth of the Mediterranean region. But on the other hand it had to do with the natural and artificially erected barriers in the east, which made it much harder to penetrate the rich lands of China. Europe simply became an easier prey – and often a fallback for those who had failed in the east.

One of the equestrian peoples that came to Europe about 1,000 years after the Scythians probably consisted of elements of the Rouran tribe. After being crushed by the Mongolian Turks, the Rouran are thought to have found their way to the west in order to establish the Avar

Khaganate. This was previously just an educated guess, but now it is supported by genetic data.

In 568 the Avars took over the Carpathian Basin in the western part of present-day Hungary. Over the following decades they consolidated their power base between the Carpathian and the Balkan Mountains to such an extent that they were able to extract tributes from the neighbouring Byzantine and Frankish empires. The Avars took with them to the grave most of the wealth they had accumulated in this way – in sharp contrast with some of the Europeans, who were by then Christianized, so that the practice of offerings for the afterlife had all but died out among them. The funeral rites of the Avars are a boon for modern archaeologists and archaeogeneticists, who can establish the status of buried individuals with great accuracy from the plethora of objects buried with them (see Figure 37). Moreover, the Avars left to posterity thousands of

Figure 37 Treasure of Nagyszentmiklós © Kunsthistorisches Museum Wien, KHM-Museumsverband

In the sixth century, members of Asian equestrian tribes founded the legendary Avar Khaganate in the region of present-day Hungary. The Treasure of Nagyszentmiklós, as it is known, is also attributed to the Avars, though its precise ownership can no longer be conclusively determined.

well-preserved skeletons, from which DNA has been successfully isolated in recent years. And it turns out that the Avars did indeed bear a genetic resemblance to people who lived much further east at that time – namely the Rourans, who are in turn closely related to the inhabitants of north-eastern China today.

This doesn't tell us definitively whether the Avars are descended from the Rourans or rather from another equestrian people of this region. What the genetic analyses prove beyond doubt, however, is the sheer speed of an expansion that put everything that came before it in the shade. Indeed, it was little short of a walkover, given that the interval between the warriors' departure from the eastern end of the Eurasian continent and their arrival in Central Europe apparently spanned a mere ten to twenty years. With their distinctive headdress, the Avars were described by the Byzantines as highly exotic-looking, and they seem to have kept themselves to themselves over generations, cultivating a kind of East Asian diaspora in the middle of Europe. But this must have applied only to the elites, as the bones from the more extravagant Avar graves were found to contain virtually pure East Asian DNA, while those from the poorer graves showed substantial admixtures from Europe. The Avar top brass, it seems, didn't mix with Europeans – unlike the ordinary folk.

Victory on the Bosporus

After the eventual defeat of the Avar Khaganate in the ninth century, the genetic trail of the Avatars was soon lost as well, dissolving into the giant European gene pool. The same goes for another significant steppe migration, which nowadays is detectable only in positively homeopathic doses, in individuals who lived just a few centuries afterwards. These traces were also found at the western end of the steppe, in the bones of Béla III, who ruled the kingdom of Hungary in the second half of the twelfth century. Béla was a direct descendant of the first Hungarian grand prince, Árpád, who had successfully settled the Magyars in the Carpathian Arc almost 300 years before. According to historical accounts, this tribe was descended from nomads from a region north of the Black Sea; and it brought with it the Finno-Ugric language family, whose modern European incarnations are chiefly spoken in Finland, Estonia, and Hungary.

This theory is also backed up by DNA analyses, at least in the case of Béla III: his Y chromosome – the gene locus passed down the male line and thus inherited from his ancestor Árpád – can be clearly traced back to Central Asia. The same cannot be said, however, of the rest of the genome, which is indistinguishable from the rest of the gene pool that existed in Central Europe at the time. Thus the proud clan of the Árpáds may have invaded Europe under the leadership of the first Grand Prince and in the wake of a large Magyar cavalry, but then evidently assimilated very quickly into the surrounding populations. The king's sons, too, appear to have married into Central European dynasties, so that the original genetic signature remained detectable only in the Y chromosome and is unlikely to have been recognizable in the features of subsequent Hungarian rulers.

In modern-day Hungary there are even fewer signs that the population could have originated from a tribe of nomadic horsemen from the steppe. Nor is there any genetic evidence to support the nationalistically charged theory, which has been frequently revisited in recent times, that Hungarians are descended from the Hun king Attila. In fact the Huns themselves were probably not so much a distinct population as an amalgam of disparate groups that invaded Eastern Europe on horseback, then established an empire in Hungary, and in the fifth century, under Atilla, threatened both parts of the Roman Empire.

Many historians suspect that the Huns had a background similar to that of the Avars: they could have emerged from the Xiongnu Empire, which was defeated at the northern border by the Han dynasty, in the first century. As with so many other steppe migrations, however, this is almost impossible to prove.

The same goes for another momentous migration whose origins are, again, hard to verify but easy to hypothesize. The roots of the modern 'Turkic peoples' – probably a loose collection of nomadic equestrians – are thought to lie somewhere between the Kazakh steppe and Manchuria in northern China. In the eleventh century, Anatolia was invaded by Turkic-speaking conquerors whose descendants went on to found the mighty Ottoman Empire. Today the Turkic language belt extends from the Bosporus to China, the Uyghurs being another product of the Turkic peoples' expansion.

Cannons and pathogens

The era of steppe migrations, which shaped Europe and lasted for several millennia, finally came to an end in the late Middle Ages – and did so in gruesome style. Having brought together various Mongol tribes in the eleventh and twelfth centuries, Genghis Khan and his descendants established a global empire that not only reached as far as present-day Ukraine but succeeded, under Genghis' grandson Kublai Khan, in ruling over China for a period of ninety years. This 'Yuan dynasty' lasted into the year 1368.

Another empire that grew out of Genghis Khan's campaign was the Golden Horde, which stretched from western Siberia to the northern coast of the Black Sea. It was along this corridor that the plague pathogen found its way to Europe in 1346. Genetic analyses show that the bubonic bacterium can be traced to a single point of entry, and there are historical accounts describing how besieging Golden Horde warriors used the corpses of plague victims as a biological weapon, catapulting them over the walls of the Black Sea city of Caffa in order to break the inhabitants' resistance. The attack succeeded, and with the ships of the fleeing inhabitants the deadliest cargo of the entire Middle Ages arrived in Central Europe, from where the plague began to spread like wildfire. Even after the devastating pandemic of the Black Death, the plague bacterium remained a fact of life in Europe and Asia. Since then, at least 7,000 outbreaks in medieval and modern times have been recorded in Europe alone, one of the most recent being that of Madagascar in 2017. All plague epidemics – including the third pandemic, which broke out in Hong Kong in the mid nineteenth century – have arisen from the same strain of bacteria that was spread with such calculated violence nearly 700 years ago. It was not until the invention of antibiotics in the early twentieth century that this disease – like many others – lost its terror (see Figure 38).

The Bronze Age and the subsequent Iron Age had turned the whole continent of Eurasia into a hotbed of weapons and warfare; but history was shaped just as much – and no less devastatingly – by viruses and bacteria. How many victims were claimed by plague, leprosy, tuberculosis, smallpox, measles, influenza, and all the other pathogens – which we are perhaps not even aware of – can no longer be reconstructed,

Figure 38 A cart of plague victims at Elliant drawn by a woman in rags. Lithograph by Jean-Pierre Moynet, 1852, after a sketch by Louis-Jean-Noël Duveau

In the fourteenth century, the Black Death spread fear and terror in Europe. It was a trauma that was to last for a very long time, sweeping across the continent, in waves, over the next four centuries.

or even guessed at. One thing is certain, though: through coevolution with pathogens, Europeans and Asians of the Neolithic period gained a battle-hardened immune system – a protective shield that Americans, Australians, and the peoples of Oceania lacked when the newcomers invaded their land.

Navigation, weapons technology, and the fateful decision of a Chinese emperor to give up his fleet may all have helped to smooth the Europeans' path to world dominance from the sixteenth century onwards. And yet one of their most potent weapons turned out to lie in biological micro-organisms and inert clusters of molecules, whose function humans are only just beginning to fully understand – and which today, in the early twenty-first century, are threatening to shake us once again to our very foundations.

10

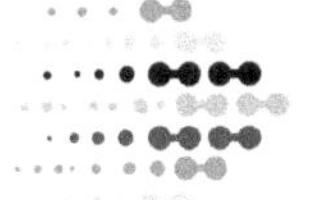

Homo hubris

The twenty-first century is destroying our self-delusion: we are not God. Did viruses make us what we are today? On the way ahead, evolution won't help us: from now on we're on our own. Looking at the Milky Way, the prospects aren't good.

An armoury of pathogens

The Black Death was a once-in-a-millennium catastrophe, and it paved the way for a new age. At least a third of Europeans – or, according to some estimates, more than a half – succumbed to the plague. The pandemic was preceded by a sustained period of growth, not least because of the changing climate brought in by the medieval warm period (MWP), which began at the end of the first millennium and lasted for about 300 years. Higher temperatures, notably in the northern hemisphere, led to increasing agricultural yields, and Greenland was settled on and off by the Vikings. Large parts of Europe saw a population explosion that produced densely inhabited and unhygienic cities, which were a breeding ground for epidemics. In these circumstances the Black Death was a disaster waiting to happen – and one towards which the Europeans, believing themselves to be on a never-ending roll, headed without an inkling of the deadly danger that loomed above their communities.

By the fourteenth century, whose middle years saw the reign of the Black Death, the warm period was gradually coming to an end. The Vikings began to withdraw from Greenland, and especially the northern latitudes of Europe, Asia, and America experienced famines that were presumably due to increasingly bad harvests. Scientists have identified

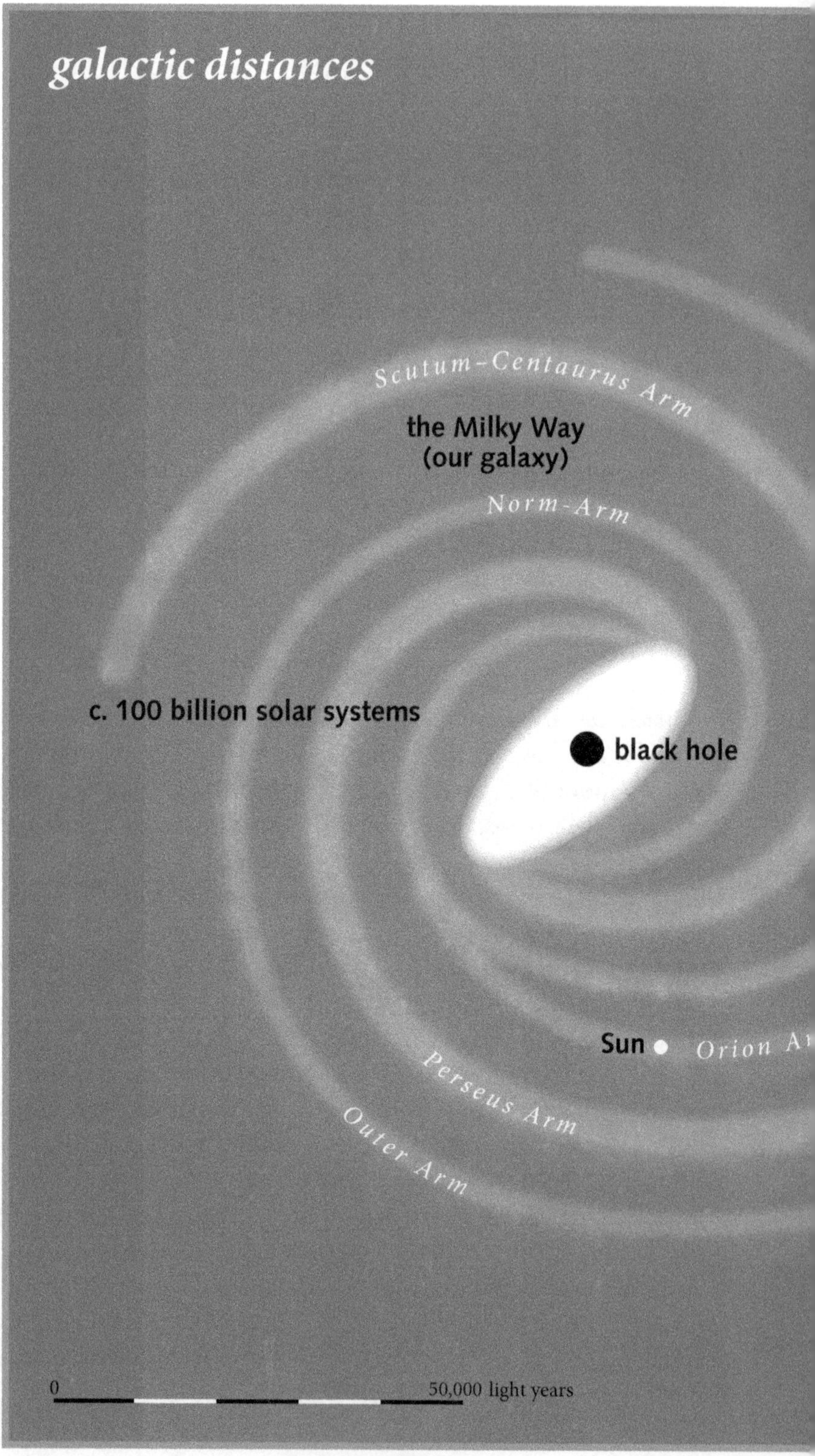

Figure 39 Galactic distances.

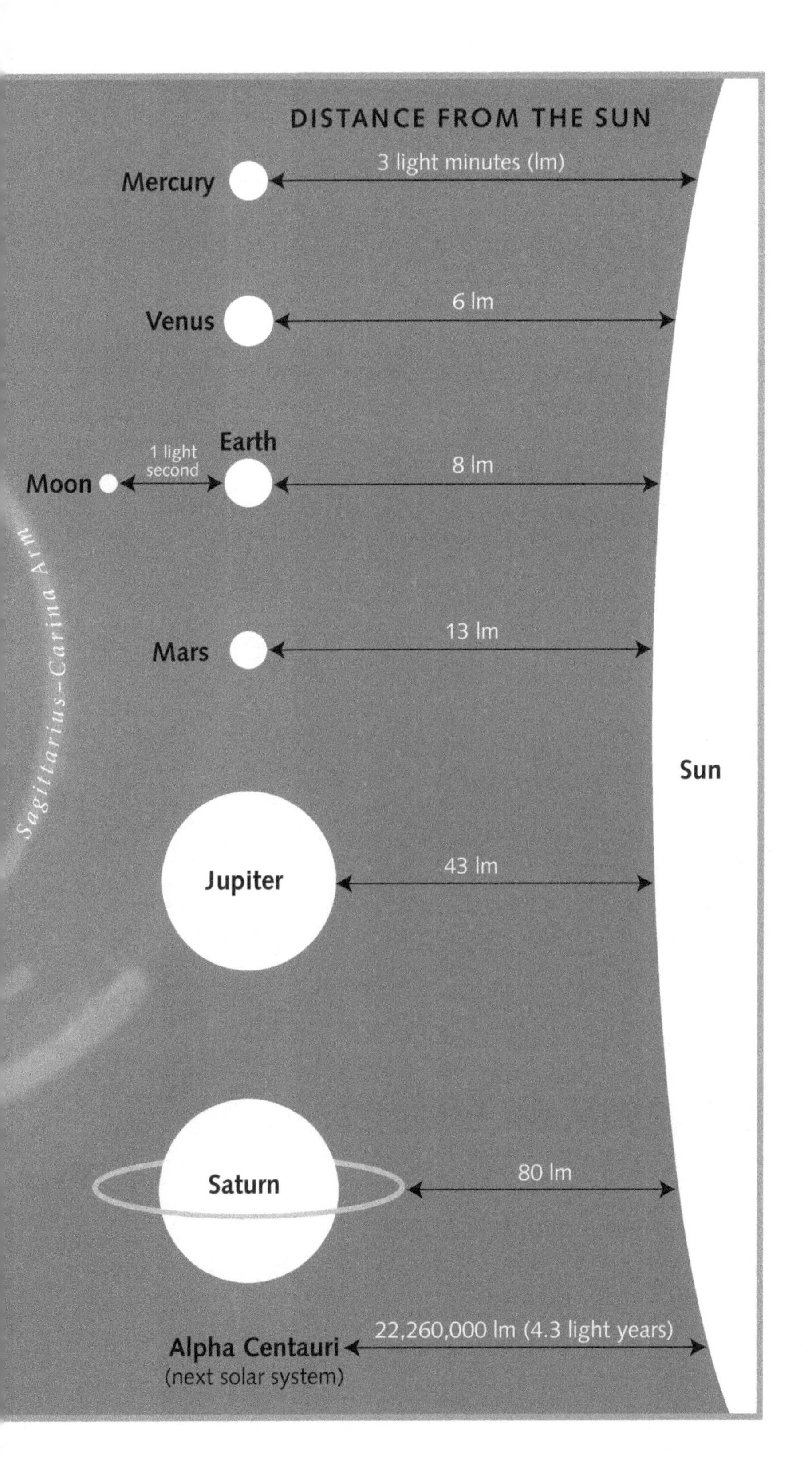
DISTANCE FROM THE SUN
Mercury
3 light minutes (lm)
Venus
6 lm
Earth
1 light second
Moon
8 lm
Mars
13 lm
Sagittarius–Carina Arm
Sun
Jupiter
43 lm
Saturn
80 lm
Alpha Centauri
(next solar system)
22,260,000 lm (4.3 light years)

several possible causes of the cooling effect, including a weakening of solar radiation and atmospheric impacts of major volcanic eruptions. The American palaeoclimatologist William Ruddiman harboured a different suspicion, though. After the pandemic, he argued, the drastically reduced population also triggered a shrinkage of the agricultural land, since less grain was needed to meet the demand. This in turn may have led to far more forestation, and hence to a reduction of carbon dioxide in the atmosphere. The Little Ice Age, which developed differently from one continent to the next, lasted into the nineteenth century, being ultimately replaced by global warming, over which humanity is now at risk of losing all control (Figure 40).

The Ruddiman theory is just one among many. But the cataclysmic effects of the plague on the course of European and hence world history

Figure 40 Hendrik Avercamp, *Winterlandschaft*, c. 1630 © National Galleries of Scotland

After a long warm period that led to the colonization of Greenland by the Vikings (among other things), the fifteenth century saw the beginning of the Little Ice Age. One theory attributes it to the interaction between humans, pathogens, and climate.

are indisputable. The dramatic reduction of the working population saw the value of labour increase after the pandemic, while a surplus of agricultural land served to lessen the oppressive power of the landowning elite. With more money in circulation, craftsmanship and trade grew in importance: cities expanded, merchants banded together in the Hanseatic League, and a self-confident middle class led medieval society into the modern age, which saw Europeans colonize the entire world. Although one shouldn't overstate the role of the plague in Europe's and Asia's highly complex and interwoven history, no one can say how the continent might have developed in the absence of such a catastrophic event.

The same goes for other pandemics. The Justinianic plague, which was similarly caused by the pathogen *Yersinia pestis* and broke out in Constantinople in 542, leaving traces as far afield as England, was a major cause of the failure of the Eastern Roman Empire's attempts to expand westward. The Bronze Age plague, too, seems very likely to have coincided with the collapse of several empires. As for the Antonine plague, which afflicted Rome in the second century, the symptoms described sound more like smallpox – nor has any DNA from plague bacteria been found yet in graves from that time. Smallpox was eradicated only through vaccination, no earlier than the 1970s, and has probably claimed more victims over the course of human history than any other disease known to us.

So far, almost all successful archaeogenetic reconstructions have been of infectious diseases caused by bacteria. They are rarely possible for viruses, where the genetic material often consists not of DNA, but of the much more unstable RNA.[1] One exception is the DNA hepatitis B virus (HBV). This microbe positively epitomizes the invariably disastrous and usually irreversible transfer of highly dangerous pathogens to the ecosystem of *Homo sapiens*. HBV has been detected in hunter-gatherers throughout the world; the oldest sample comes from South America, from an individual who died 12,000 years ago. Pedigree analyses show that the virus probably jumped from Southeast Asian Old World monkeys to humans about 20,000 years ago, presumably through blood contact or through the consumption of undercooked monkey meat – a classic case of zoonosis. Since then, the virus has continued to spread across the globe, so that over 2 billion people have been infected or are currently infected with HBV to date. About 1 million die every year from the infection, which leads to chronic illnesses but can also trigger acute and often fatal liver inflammations.[2]

Our knowledge of the history of human epidemics and pandemics is undoubtedly limited to the cruellest cases. This applies particularly to the warmer, more humid regions of the world, where many pathogens thrive but also break down more easily and are thus almost impossible for archaeogeneticists to detect. Over the past millennia all kinds of bacteria and viruses have flourished, not only in Eurasia but also in Africa, where human immune systems have had even more time to adapt to these pathogens. For the modern European conquerors of America, these bugs were like an armoured vehicle that enabled them to bulldoze their way across the double continent: the Indigenous people were all but defenceless against them.

The typhus bacterium, for example – discovered in 2018 in a southern Mexican cemetery and introduced by the Europeans – probably killed around half of the population of Central America during the Great Cocolitztli Epidemic of the mid sixteenth century. Infectious diseases were brought to America from Africa too, and then they spread to the local people. In 2020 the bacteria responsible for yaws were found in the bones of African slaves trafficked to America by Europeans. Like syphilis, this pathogen often lies dormant in the body for decades until a weakened immune system allows it to do its deadly work.

This highly developed immune response was common to both Europeans and Africans, though their circumstances couldn't have been more different. After they had not only enslaved the South Americans but almost wiped them out with their diseases, the Europeans set their sights on the African slaves, whom they saw as more suitable because they were more resistant: in other words, their innate immune system was already adapted to Old World pathogens. This, however, did not make the African slaves' encounter with the new rulers of the world any less fatal.

Preparing for the next pandemic

Over the past decades the destructive power of pathogens had been gradually fading from the memory of the global West, while remaining an ever-present danger for people in lower-income countries. The recent Covid-19 pandemic has now brought the power of these microbes sharply back into focus for all of us.

Covid-19 cannot, however, be compared with the Black Death. Even so, we can't rule out the possibility of a similarly deadly pathogen spreading around the globe. Ebola has already reared its head, and could be brought under control only because the African continent is less involved in the intensive global movement of goods and people than the northern hemisphere. Factory farming, with its excessive use of antibiotics, has become a breeding ground for more and more antibiotic-resistant microbes. These already claim an estimated annual death toll of around 700,000, of which 200,000 – and the number is rising – can be attributed to resistant tuberculosis strains. If a highly infectious and deadly pathogen were to evolve in this slaughter ground that serves the human hunger for meat – and we are currently doing everything possible to help it along – we could see the emergence of a pandemic that would put Covid in the shade.

Going by experience, however, and with a little trust in the development of new antibiotics that are not immediately blunted by agriculture, the next pandemic will come from a virus. The biggest threat here is probably from novel influenza pathogens. H1N1, for example, was transmitted to humans during the early twentieth-century outbreak of Spanish flu and generated a pandemic that killed far more people than the First World War just before. In 2005, H1N1 became the first virus from the past to have its genome sequenced. It was also responsible for the swine flu pandemic of 2008 and 2009. One reason for the relative mildness of the later episode may have been that many older people had already come into contact with the same strain in their youth (the Spanish flu pathogen was probably still circulating into the 1960s), and had an immune response at the ready. Had swine flu developed the same kind of momentum that Spanish flu once did – and today's much increased human mobility and population size offered ideal conditions for this – we would have seen hundreds of millions of deaths worldwide. In the event, though, the maximum death toll was 'only' 500,000.

The Neandertal: An evolutionary blunder?

Given the countless numbers of people killed by viruses over the course of human history, it may seem absurd to regard these inert bundles of molecules as any kind of force for good. But the opposite is true. Without viruses, human evolution would probably never have happened – or at least not at the dizzying speed we have seen. For viruses don't just infect organisms in order to exploit the host's replication mechanisms for their own reproductive ends: they can also become part of the body's DNA. A case in point is HIV: this retrovirus worms its way into the genome of a human cell and alters the blueprint so that the body becomes a self-directed munition factory.[3] In the course of evolution, though, other viruses have lost the ability to form a capsid or protein shell, which means that they can only copy their own genetic material and insert it back into the human genome; such genomic viruses are also known as 'retrotransposons'. The DNA of almost all living things is imbued with them: indeed, almost half of our genome derives from retrotransposons. And that's no bad thing.

Here's why: retrotransposons have the ability to speed up evolution: if these genomic viruses get into the chromosomes, this can trigger a whole cascade of events. Since DNA is constantly duplicating and recombining with itself, in the best-case scenario the intrusion of new base sequences acts like a trump card. Basically the same rule that applies to population sizes works for the genome too: the more variance, the greater the advantage in the evolutionary race.

Birds, for example, are a class with a relatively low proportion of retrotransposons: it accounts only for a tenth of their genome. While this doesn't prevent them from dominating the skies, they also exhibit a remarkable genetic constancy, all the way back to the archaeopteryx, the earliest known bird. Although bird chromosomes have changed over the past 150 million years, their basic structure remains largely the same. With us great apes, it's a very different matter. In our case, not only is the proportion of genomic viruses higher than in other animals, but they are also dominated by a group called ALU (*Arthrobacter luteus*) retrotransposons, which act like turbochargers, serving to fast-track the stages of evolution. Although all primates have these ALUs in their genome, new ones have spread in Old World monkeys over the past 40 million years. This could be one reason why the evolution from monkey to human took place in Africa, and not somewhere else with an equally favourable climate – such as South America, the home of New World monkeys.

That said, high genomic variation isn't necessarily a passport to evolutionary glory. The development of ever new genetic characteristics is a good thing only if they chime with natural environment. For the *Homo* genus, that was indeed the case: our ancestors' rapid evolution and expansion northwards took place in the Ice Age – a phase of intense climatic variation and extremes that required maximum adaptability. We know this because we are at the end of that career ladder. But all the lineages that were spawned by the great ape family over millions of years and that have all since disappeared, be it in Africa or, as with 'Udo', in the foothills of the Alps – go to show the vanishingly small chances of evolutionary branches to survive in the long term. During the long time that has elapsed since the common ancestors of chimpanzees, bonobos and modern humans were alive, only three representatives have succeeded: chimpanzees, bonobos, and modern humans.

In this case, less was more – or so one could argue, given the 'up like a rocket, down like a stick' history of the Neandertals. After all, if you consider the ancestors we share with the Neandertals, you can't help but conclude that we are genetically much closer to *Homo erectus* than to the Neandertal. Physiognomically speaking, this hominin – with his conspicuously flat head, prominent brow ridge, large nose, and wide paranasal sinuses – had gone down an evolutionary path that departed much further from the original than our own. During the Ice Age, it must have been these very mutations, which occurred in quick succession, that enabled our cousins to advance in the cold north. There a robust physique with a low ratio of surface area to body mass helped to reduce heat loss through the skin.

The evolutionary sprint performed by the Neandertals of Ice Age Europe can be compared to the act of someone enthusiastically digging an ever deeper tunnel in the perpetual ice. No one could beat them at that game. But when it came to the adaptability and flexibility needed to conquer new territories and hunting grounds in all directions, the ice-bound Neandertal was left high and dry. Evolution had made a blunder.

The pandemic force of humanity

And yet, as we know, it took hundreds of thousands of years and many false starts before our ancestors were able to take the place of the Neandertals and Denisovans. The fact that they eventually succeeded, even though there had been no real change in the extreme conditions of the north, shows that biological factors must have been at play. How else could modern humans have achieved, in what was the blink of an eye on the scale of world history, something that had eluded the Neandertals and Denisovans for so long – as well as all the earlier modern humans who had tried and failed before them?

There must have been something in the genes of the later, successful migrants that wasn't there before. There are clear signs of this at the end of the Ice Age, when the dawn of the Holocene was followed almost immediately by the mutually independent yet basically identical development of agriculture in various corners of the world. It took humans less than 1,000 years to begin exploiting the new climatic conditions in order to plough fields and domesticate animals in the Middle East. Had this happened only in one place, it might have been a lucky accident. But the same thing happened again soon afterwards, at the other end of Asia, and a little later in America and Africa; and there may have been an autonomous Neolithic revolution in India and New Guinea too. Knowledge of plant cultivation was certainly not embedded in the human genes – but the ability to have this eureka moment in the first place was. The necessary intellectual capacity seems to have been shared by all humans, from Africa to America. So why didn't it happen before?

Remember: the Ice Age was interrupted by the Eemian interglacial period 126,000 years ago. This lasted for a good 11,000 years – a period equivalent to the interval between the beginning of the Holocene and the present day. Modern humans were probably already living in the Middle East throughout the Eemian. During that period, the global average temperature rose higher than ever before in the Holocene, and the region of present-day Israel must have offered ideal conditions for agriculture. Yet no change of that nature happened for 11,000 years: people remained true to their hunter-gatherer lifestyle. The idea of planting a seed in the ground and waiting to see what transpired never entered their heads.

Had it done so, the unstoppable rise of humanity might have begun much earlier, and we could have been living today not in the twenty-first century, but in the hundred and fifteenth. Then again, we could just as easily have continued roaming through the land as foragers to this day, spoilt by the rich bounty of the interglacial hunting fields. That is, if we still hadn't grasped, even after 11,000 years, what the temperature rise meant in terms of new opportunities for human civilization. As it is, we obviously had by then the genetic blueprint that allowed us to do exactly that.

It may have been just a few mutations in our genome, a few happy coincidences in key parts of our DNA sequence. Yet this was a shift of seismic importance for the future, not just of our species but of the entire global ecosystem. From now on, humans became a rampant force in the world. Anything that stood in their way or that it served them to exploit was treated with lethal violence. The harmless great ape from Africa had itself mutated into a pandemic force.

The vain search for the human self

For the megafauna of the non-African world, the arrival of humans invariably signalled the end. What was left of the local wildlife either had to adapt to the new rulers or was fed into an evolutionary bottleneck, to emerge at the other end as human-serving domestic animals. Humanity has penetrated the deepest regions of the earth and its oceans, and is in the process of burning through the legacy of millions of years' worth of forests and organisms in the form of coal and oil. The geological epoch that has seen us transform our planet and its protective layer, in some cases irreversibly, is nowadays referred to as 'the Anthropocene': the age of the human. It is about to replace the Holocene, the natural warm period currently still in progress. While the evolution of the world's myriad organisms used to be shaped by external conditions, it is now a single species that is changing the face of the earth irrevocably. As in a nuclear chain reaction that begins with the fusion of two atoms and, in the worst-case scenario, leads to the release of uncontrollable forces, the interplay of random genetic shifts in our ancestors generated an evolutionary momentum that bulldozed everything in its path – and thereafter prevented anything else from emerging – anything that might evade its domination (see Figure 41).

The belief in our own invincibility rose to a new level in the Neolithic period and reached new heights with the dawn of the Bronze Age – the beginning of wealth, patriarchy, and military subjugation. The only natural power that humanity was almost powerless to resist was one that it failed to recognize. In their ignorance, humans declared deadly diseases and epidemics divine punishments, and hence the instrument of a being of their own creation and imagination. In so doing, they were defending themselves against a fact that the earliest cultures had likewise sought to defy when they buried their dead and sent them to their afterlife with grave gifts: the knowledge that they, too, were part of the natural cycle, merely one animal among many, at the mercy of their environment, and liable, if they were unlucky, to be carried off by an invisible foe.

In the first half of the twentieth century, with the development of vaccines and antibiotics, humanity entered a new phase, in which it believed itself to have conquered even the seemingly divine power of pathogens. Thanks to the establishment of the concept of hygiene in everyday life and in medicine, along with rapid strides in pharmacology, modern humans finally emancipated themselves from nature – or so they thought. For many people, infectious diseases much feared in earlier periods became a relic of the past. Medical research began to focus increasingly on combating the diseases of civilization: in short, the consequences of overabundant food, less physically strenuous work, and consumption of toxic stimulants.

The decoding of the human blueprint and the emerging ability to repair it with genetic scissors or, in extreme cases, to change it for good are the latest achievements of a medical industry that has declared new breakthroughs at ever shorter intervals since the last century. All these are aimed at pushing the boundaries of the human organism to the limit. Genome sequencing has become an indispensable part of the fight against disease, and there is no doubt that it will soon be used for the further purpose of optimizing life. Evolution is too important, it seems, to be left to its own devices.

The twentieth century turned *Homo sapiens* into *Homo hubris*. Members of the new species know better than ever before how to make use of their intellect – the organ that sets them apart from all other life forms. It was this organ that got them where they are today, enabling them to shape and exploit the earth to the limits of its capacity and

Figure 41 The atom bomb © Charles Levy, US National Archives and Records Administration

The development of the atom bomb put an end to the period of the great wars. At the same time, its use would render large parts of the world inhabitable in a matter of hours and probably usher in the end of human civilization.

beyond, so as to further their own interests. Now that they have come up against planetary boundaries, their evolutionary conditioning is no longer of any use. Expansion, the constant 'onwards and upwards', is no longer an option. We recognize this, wise species that we are. But recognition is not enough.

Our rise to the position of dominant species was accompanied by the elimination of all competition: since the Neolithic period, the survival of our civilization has been bound up with the activity of asserting ourselves over other living creatures, suppressing them or destroying them, in the fight for resources. It may well be that, ever since we fashioned the first hand-axe from a stone, ignited the first fire, or painted the first animal on a cave wall, we were animated by the notion that our existence fulfils some kind of higher purpose – one that justifies us and thereby gives us a guarantee of eternity. At bottom, though, we remain to this day a product of the classic predator–prey relationship that determines almost all evolutionary processes and from which humans have up to this point emerged victorious. Now that our dominion spans the entire planet, we find ourselves face to face with the final enemy: ourselves.

Evolution is an inescapable process in which the winners are those who can best adapt to their environment. Dinosaurs existed for 150 million years, but it wasn't necessarily the smartest ones that survived. Rather it was those with the sharpest teeth or the best fight-or-flight response, and those whose plumage enabled them to ride out the catastrophe of mass extinction and either delight us today in spring and summer with their song or end up as poultry on our dinner plates. Tortoises and other reptiles haven't gained in intelligence throughout their evolution, but have developed thicker carapaces or better hunting methods. Evolutionary unique selling points are as diverse as the planet's flora and fauna itself, and human intelligence is just one of them. Crucially, however, it is one with as much deadly force as the bite of a Tyrannosaurus. Our brain is not just our answer to a dinosaur tooth: it is threatening to become the equivalent of the meteorite that wiped out most life on earth 66 million years ago, dinosaurs and all.

This is at once scary and fascinating. The 'neandertalization' of brains in a modern laboratory is simply another attempt to unlock the secret of this curiously self-destructive organ. In recent history, scientists have observed, opened up, dissected, scanned, and analysed down to the last cell innumerable specimens, yet no one has found such a thing as soul; no one has succeeded in localizing the site of human consciousness; no cultural or religious gene has been identified. And even now, when we know which DNA loci distinguish modern humans from their closest relatives, the Neandertals, when we can see it with our own eyes, in clearly defined base sequences, we still haven't managed to translate those data sequences into an answer to the question of what makes us human. Nowadays physicists can explain what happened in the first millisecond after the Big Bang, yet the core of our existence remains a mystery to us. Like elementary particles orbiting the 27-kilometre-long particle accelerator at the CERN (Conseil européen pour la Recherche nucléaire) centre in Switzerland, humanity has, from the word go, circled around the question of what makes us what we are.

An intriguing thought

We have long reached the limits of what our high-powered brain can do by itself. Indeed, it was only by outsourcing the necessary workload that we were able to decode our DNA in the first place. The task of reading genetic data and, above all, distilling the relevant information from it is performed by computers and machines that can scan, store, and link infinitely more data than us and present them to us ready-packaged for interpretation. Without big data, the discoveries made by archaeogenetics over the past two decades would be unthinkable; and the same goes for other disciplines, such as physics or medicine. The invention of the computer could turn out to be as pivotal to the future development of humanity as that of the hand axe, or as the ability to kindle fire.

What we feed into computers consists essentially of 'discrete' data – that is, 0s and 1s, clear information, yes and no, plus and minus. But when it comes to the dichotomies that all humans think they can grasp intuitively – good and evil, right and wrong, reason and unreason, progress and self-destruction – so far no machine has been able to deliver. Who knows, perhaps one day big data and artificial intelligence will not limit themselves to solving scientific problems but will answer the biggest question of our time. Perhaps AI will have the capacity to tell us who we are. And who we ought to become. An intriguing thought – and one that artificial intelligence would calculate with cold precision.

No, AI is unlikely to do that – and it doesn't need to, either. For, while the tasks that lie ahead of us may be all but insoluble from our present perspective, they are at the same time as trivial as the logic of evolution. We know that we consume many times more resources than we should. We read, again and again, that we fly too much, eat too much meat, dispose of too much plastic, fell too many forests, seal too much land, pollute too much ground water, consume far too much. Understanding is not the problem. The problem is our deep-seated aversion to submitting to natural limitations, particularly when they are not yet palpable, in the here and now, but only abstractions. We cannot get out of our skin.

At the same time, we know that it would be possible to feed a world population of more than 10 billion, and even enable them all to live a decent life. Such a world would be easy to design on the drawing board: we would simply have to divide the resources available within the planetary boundaries by the number of people who use them, taking into account the regeneration processes ensured by natural cycles. We probably wouldn't even need a computer to do it. But all this runs counter to the biology of a creature that owes its unique success story to the incredibly competitive organ between its ears. Equitable distribution is not in our nature. Nor is it in our culture, whose logic of 'bigger, better, faster' is not just a consequence of our genetic development but its whole premise.

Alone in space

Homo hubris appears to be incapable of accepting the limits of growth. This is reflected not just in our relentless competition for resources but also in the fact that, having conquered Earth, we are now setting our sights on the cosmos. Ever since the first human flew into space, only to be trumped shortly afterwards by the first person to land on the moon, we have cherished the dream of expanding human civilization into the wider universe. At the beginning of the millennium, Elon Musk founded the aerospace company SpaceX, with the declared aim of conquering Mars; the billionaires Jeff Bezos and Richard Branson then later joined the space race and were both blasted into the cosmos for a few brief minutes in 2021. That there would be enough takers for Elon Musk's envisaged mission to a planet 55 million kilometres away whose inhospitable environment holds nothing that bears even the vaguest resemblance to life on Earth is beyond question, if we look at human history so far. What is much more doubtful is that this could be a plan B in case we destroy the basis for life on Earth. After all, if it were possible for an intelligent species to conquer space, we would very likely know about it by now.

This idea was first expressed in 1950 by the Nobel Prize-winning physicist Enrico Fermi, in a now legendary casual conversation that was later distilled into the so-called Fermi paradox. The paradox is based on the assumption that, since the Big Bang life must have formed throughout the universe – not only on Earth but presumably in countless millions of cases. According to current calculations, in the tiny speck of the universe that makes up the Milky Way alone there are at least 100 billion stars, and perhaps even up to 400 billion. Around 5 to 20 per cent of these resemble our sun, and some of these suns are probably orbited in turn by planets with life-friendly conditions, just like those on Earth.

According to Fermi, intelligent life must have formed hundreds, perhaps even millions of times over since the beginning of our 13.6-billion-year-old galaxy. And every intelligent life form – so the theory goes – will, at some stage in its evolution, first reach the point where it has exhausted its planetary boundaries and, second, develop space travel. It then follows, as inevitably as a law of nature, that it will try to colonize other solar systems – just as humanity appears to be doing even as we speak. Since Earth is probably one of the younger planets of the Milky Way at the age of barely 4.6 billion years, there would have to have been innumerable previous instances of aliens spreading out from their home planet and eventually winding up, like the island hoppers of the Austronesian expansion, on our positively exemplary Earth – a veritable El Dorado for carbon-based life forms.

This, in a nutshell, is the scary anomaly described by the Fermi paradox. On the one hand, the existence of myriad intelligent life forms in the Milky Way is a virtually inescapable logical conclusion. At least some of these would have had time over the past few billions of years to colonize the entire galaxy several times over – and would indeed have done so. On the other hand, there are no traces of aliens on Earth, and the main argument against their existence is that, during all this time, we alone have had the evolutionary run of the place. Nor has the exploration of our galactic neighbourhood offered the slightest indication of there being intelligent life somewhere out there – at least not of a kind that sends out signals or launches satellites into space, as we do. That we are and always have been the only intelligent life form in a galaxy that spans a few hundred thousand light years is one of the mathematically most unlikely conclusions. There is of course another possible explanation though, and a chilling one at that: there may have been a succession of intelligent life forms in the Milky Way, but each one followed the same rapid evolutionary path as we did – and none is still around today (see Figure 42).

Thirteen billion years, billions of suns, an unimaginable number of successively flourishing civilizations – yet apparently not one that has managed to put out feelers into space. Are we the latest addition

Figure 42 Capturing the Milky Way © ESO / B. Tafreshi (twanight.org)

Gazing up into our galaxy has been a humbling experience for humans ever since they became capable of thought. Now *Homo hubris* dreams of conquering the Milky Way. But, for the first time in our history, our future doesn't lie in expansion. It is time to reinvent ourselves.

to an ancestral gallery of failed Milky Way civilizations? Just another species that thinks it doesn't have to accept limits? And is our decline as inevitable as the Neolithic decline into which *Homo sapiens* found itself sleepwalking? Or perhaps the other intelligent species of our galaxy didn't disappear after all and are even now observing the next experimental set-up, fascinated by our inability to grasp the necessary next step on our evolutionary journey – namely towards a lifestyle in which planetary boundaries become an integral part of our consciousness.

A near-perfect blueprint

The fate of humanity will be decided in the twenty-first century. The natural resources of our planet will soon be exhausted. We will all be obliged to draw on the same limited pool. The potential for deadly conflicts to spread quickly round the globe will increase in the years ahead: competition for resources, trade routes and spheres of influence is already dominating international relations. At the same time, the world is armed with at least 13,000 nuclear warheads, and the almost inevitable chain reaction that their use would trigger hangs on a push of a button by an irrational ruler. In 2020 the 'doomsday clock' was set at 100 seconds to 12, the highest level since its invention in 1947.[4] If climate change and the dawning age of the pandemic are the dark clouds on the horizon, then the nuclear arms race is a cocked revolver we are currently holding to our heads. And that danger is set to increase the more existential global resource conflicts become: the threat of wars over drinking water in the regions most afflicted by a warming climate is a case in point.

The Covid-19 pandemic has shown up, as if under a magnifying glass, how much the fate of all human beings is interconnected. During the crisis, the last resort of a globalized world was to close borders, albeit at a cost that is difficult to quantify. Come the next crisis – a much more severe one – that last resort won't be possible. How will the world deal with states that continue to place a disproportionate burden on the climate, thereby increasing the risk to us all? How should democratic societies meet the challenge of reconciling the cherished freedom to consume with nature's shrinking budget forecast for the coming decades? These are questions to which very real answers need to be found, and soon – through government programmes, laws, and international agreements backed up by effective sanctions.

The lasting peace of the second half of the twentieth and early twenty-first century that a large part of the global population can look back on today – and that has already started crumbling in Eastern Europe and the Middle East – was a unique civilizational accomplishment, perhaps the greatest cultural achievement our species has proved capable of so far; and yet it is based on a deal done to the detriment of others and, ultimately, to our own detriment. No one knows how a being programmed by its DNA for expansion, consumption, and subjugation will cope with the problem of dwindling resources, with the challenge of tightening its belt. The call to cut down – on flying, on meat, on cubic capacity – is already a recipe for deep social division. And such conflicts will become all the more dangerous in a highly armed world soon to be dominated by multiple centres of power.

The idea that everything we hold dear is at stake is one we refuse to accept. And yet it doesn't take much hindsight to see that our unconditional belief in our superiority is a fallacy. The improbable coincidences of our evolution, the many setbacks – all this appears to us today like one steady upward curve. What we don't see is the human lineages that ended in volcanic ash, in the sea, or in an obscure corner of Africa or Eurasia. We forget that our ancestors had to pass through countless evolutionary bottlenecks; that pandemics, climate disasters, and wars have persistently annihilated large sections of entire populations. All of us living in the world today are descendants of survivors. And just as there was no survival guarantee for the many defunct lineages of modern humans, there is none for us either.

This is, in spite of everything, an encouraging thought.

After all, there was nothing automatic about our ancestors' evolution, and the road to perdition is not a foregone conclusion either. Who, if not *Homo hubris*, is up to the task of protecting us from ourselves? Our natural evolution has long since ended and will take us nowhere: there are simply too many people on the planet for a mutation in any one individual, however beneficial, to spread throughout the population in a short space of time. It is now up to us to manage our near-perfect blueprint in such a way as to avoid triggering its intrinsic self-destruction mechanism. The salvation of human civilization will surely be a cultural achievement – and one that our descendants may one day speak of with the same reverence that we reserve for the first cave paintings of Stone Age humans.

Our ancestors came to conquer. They came to stay. Something about them was different. Exactly what that something is, we may never fully understand. All we know is that we hit the evolutionary jackpot, perhaps as the only species to do so in the Milky Way, on a planet that is quite unique. The challenge now is to avoid frittering away our winnings. It is time for the next great leap: to a world with which we are content.

Sources

For easier reading, we have not included footnotes for the source references in this book. The following is a list of publications – journal articles, books and other sources – that were consulted for the individual chapters. They are not listed in alphabetical order. In addition, some statements in the book are based on conversations with fellow researchers whose assessments and interpretations, where shared by the authors, have been included in the text. Multiple mentions have been avoided.

Chapter 1

Green, R. E. et al., A draft sequence of the Neandertal genome. *Science* 2010, ***328***(5979): 710–22.

Reich, D. et al., Genetic history of an archaic hominin group from Denisova Cave in Siberia. *Nature* 2010, ***468***(7327): 1053–60.

Meyer, M. et al., A high-coverage genome sequence from an archaic Denisovan individual. *Science* 2012, ***338***(6104): 222–6.

Prufer, K. et al., The complete genome sequence of a Neanderthal from the Altai Mountains. *Nature* 2014, ***505***(7481): 43–9.

Prufer, K. et al., A high-coverage Neandertal genome from Vindija Cave in Croatia. *Science* 2017, ***358***(6363): 655–8.

Slon, V. et al., A fourth Denisovan individual. *Science Advances* 2017, ***3***(7): e1700186.

Slon, V. et al., The genome of the offspring of a Neanderthal mother and a Denisovan father. *Nature* 2018, ***561***(7721): 113–16.

Bokelmann, L. et al., A genetic analysis of the Gibraltar Neanderthals. *PNAS* 2019, ***116***(31): 15610–5.

Peyregne, S. et al., Nuclear DNA from two early Neandertals reveals 80,000 years of genetic continuity in Europe. *Science Advances* 2019, ***5***(6): eaaw5873.

Krause, J. et al., The complete mitochondrial DNA genome of an unknown hominin from southern Siberia. *Nature* 2010, ***464***(7290): 894–7.

Pääbo, S., *Neanderthal man: In search of lost genomes*. New York: Basic Books, 2014.

Reich, D., *Who we are and how we got here: Ancient DNA revolution and the new science of the human past*. New York: Pantheon Books, 2018.

Prufer, K. et al., The bonobo genome compared with the chimpanzee and human genomes. *Nature* 2012, ***486***(7404): 527–31.

Meyer, M. et al., Nuclear DNA sequences from the Middle Pleistocene Sima de los Huesos hominins. *Nature* 2016, ***531***(7595): 504–7.

Stepanova, V. et al., Reduced purine biosynthesis in humans after their divergence from Neandertals. *eLife* 2021, ***10***: e58741.

Lancaster, M. A. et al., Cerebral organoids model human brain development and microcephaly. *Nature* 2013, ***501***(7467): 373–9.

Jinek, M. et al., A programmable dual-RNA-guided DNA endonuclease in adaptive bacterial immunity. *Science* 2012, ***337***(6096): 816–21.

Greely, H. T., Human germline genome editing: An assessment. *CRISPR Journal* 2019, ***2***(5): 253–65.

Cyranoski, D., Russian biologist plans more CRISPR-edited babies. *Nature* 2019, ***570***(7760): 145–6.

Racimo, F. et al., Archaic adaptive introgression in TBX15/WARS2. *Molecular Biology and Evolution* 2017, ***34***(3): 509–24.

Sankararaman, S. et al., The combined landscape of Denisovan and Neanderthal ancestry in present-day humans. *Current Biology* 2016, ***26***(9): 1241–7.

Dannemann, M., and J. Kelso, The contribution of Neanderthals to phenotypic variation in modern humans. *American Journal of Human Genetics* 2017, ***101***(4): 578–89.

Dannemann, M., K. Prufer, and J. Kelso, Functional implications of Neandertal introgression in modern humans. *Genome Biology* 2017, ***18***(1): 61.

International Human Genome Sequencing Consortium, Finishing the euchromatic sequence of the human genome. *Nature* 2004, ***431***(7011): 931–45.

Gansauge, M. T., and M. Meyer, Single-stranded DNA library preparation for the sequencing of ancient or damaged DNA. *Nature Protocols* 2013, ***8***(4): 737–48.

Neubauer, S., J. J. Hublin, and P. Gunz, The evolution of modern human brain shape. *Science Advances* 2018, ***4***(1): eaao5961.

Gunz, P. et al., Neandertal introgression sheds light on modern human endocranial globularity. *Current Biology* 2019, ***29***(1): 120–7, e5.

Church, G. M., and E. Regis, *Regenesis: How synthetic biology will reinvent nature and ourselves*. New York: Basic Books, 2012.

Bethge, P., and J. Grolle, Interview with George Church: Can Neanderthals be brought back from the dead? *Der Spiegel*, 21 January 2013.

Chapter 2

Brunel, E., J.-M. Chauvet, and C. Hillaire, *Die Entdeckung der Höhle Chauvet-Pont d'Arc*. Saint-Rémy-de-Provence: Editions Equinoxe, 2014.

Conard, N. J., M. Malina, and S. C. Munzel, New flutes document the earliest musical tradition in southwestern Germany. *Nature* 2009, ***460***(7256): 737–40.

Rosas, A. et al., Paleobiology and comparative morphology of a late Neandertal sample from El Sidron, Asturias, Spain. *PNAS* 2006, ***103***(51): 19266–71.

Lalueza-Fox, C. et al., Mitochondrial DNA of an Iberian Neandertal suggests a population affinity with other European Neandertals. *Current Biology* 2006, ***16***(16): R629–30.

Yustos, M., and J. Y. Sainz de los Terreros, Cannibalism in the Neanderthal World: An exhaustive revision. *Journal of Taphonomy* 2015, ***13***(1): 33–52.

Rougier, H. et al., Neandertal cannibalism and Neandertal bones used as tools in Northern Europe. *Scientific Reports* 2016, ***6***: 29005. doi: 10.1038/srep29005.

Parrado, N., and V. Rause, *72 Tage in der Hölle: Wie ich den Absturz in den Anden überlebte*. Munich: Goldmann Verlag, 2008.

Berger, T. D., and E. Trinkaus, Patterns of trauma among the Neandertals. *Journal of Archaeological Science* 1995, ***22***: 841–52.

Schultz, M., Results of the anatomical–palaeopathological investigations on the Neanderthal skeleton from the Kleine Feldhofer Grotte (1856) including the new discoveries from 1997/2000. *Rheinische Ausgrabungen* 2006, ***58***: 277–318.

Xing, S. et al., Middle Pleistocene human femoral diaphysis from Hualongdong, Anhui Province, China. *American Journal of Physical Anthropology* 2021, ***174***(2): 285–98.

Les Abbés, A., J. Bouyssonie, and L. Bardon, Découverte d'un squelette humain moustérien à La Bouffia de la Chapelle-aux-Saints. *L'Anthropologie* 1909, ***19***: 513–8.

Prufer, K. et al., A genome sequence from a modern human skull over 45,000 years old from Zlatý kůň in Czechia. *Nature Ecology & Evolution* 2021, ***5***(6): 820–5.

Vlček, E., The Pleistocene man from the Zlatý kůň cave near Koněprusy. *Anthropozoikum* 1957, ***6***: 283–311.

Prošek, F., The excavation of the 'Zlatý kůň' cave in Bohemia: The report for the 1st research period of 1951. *Československý kras* 1952, ***5***: 161–79.

Chapter 3

Scerri, E. M. L. et al., Did our species evolve in subdivided populations across Africa, and why does it matter? *Trends in Ecology and Evolution* 2018, ***33***(8): 582–94.

Sankararaman, S. et al., The genomic landscape of Neanderthal ancestry in present-day humans. *Nature* 2014, ***507***(7492): 354–7.

Grun, R., and C. Stringer, Tabun revisited: Revised ESR chronology and new ESR and U series analyses of dental material from Tabun C1. *Journal of Human Evolution* 2000, ***39***(6): 601–12.

Hershkovitz, I. et al., Levantine cranium from Manot Cave (Israel) foreshadows the first European modern humans. *Nature* 2015, ***520***(7546): 216–9.

Meyer, M. et al., Nuclear DNA sequences from the Middle Pleistocene Sima de los Huesos hominins. *Nature* 2016, ***531***(7595): 504–7.

Posth, C. et al., Deeply divergent archaic mitochondrial genome provides lower time boundary for African gene flow into Neanderthals. *Nature Communications* 2017, ***8***: 16046.

Petr, M. et al., The evolutionary history of Neanderthal and Denisovan Y chromosomes. *Science* 2020, ***369***(6511): 1653–6.

Harvati, K. et al., Apidima Cave fossils provide earliest evidence of *Homo sapiens* in Eurasia. *Nature* 2019, ***571***(7766): 500–4.

Hershkovitz, I. et al., The earliest modern humans outside Africa. *Science* 2018, ***359***(6374): 456–9.

Harting, P., Le système éemien. *Archives néerlandaises des sciences exactes et naturelles de la Societé Hollandaise des Sciences (Harlem)* 1875, ***10***: 443–54.

Preece, R. C., Differentiation of the British late Middle Pleistocene interglacials: The evidence from mammalian biostratigraphy. *Quaternary Science Reviews* 1999, ***20***(16–17): 1693–705.

Besenbacher, S. et al., Direct estimation of mutations in great apes reconciles phylogenetic dating. *Nature Ecology & Evolution* 2019, ***3***(2): 286–92.

Bohme, M. et al., A new Miocene ape and locomotion in the ancestor of great apes and humans. *Nature* 2019, ***575***(7783): 489–93.

Scally, A. et al., Insights into hominid evolution from the gorilla genome sequence. *Nature* 2012, ***483***(7388): 169–75.

Tenesa, A. et al., Recent human effective population size estimated from linkage disequilibrium. *Genome Research* 2007, ***17***(4): 520–6.

Swisher, C. C., G. H. Curtis, and R. Lewin, *Java Man: How two geologists changed our understanding of human evolution.* Chicago, IL: University of Chicago Press, 2002.

Gargani, J., and C. Rigollet, Mediterranean Sea level variations during the Messinian Salinity Crisis. *Geophysical Research Letters* 2007, ***34***(10).

Lieberman, D., *The story of the human body: Evolution, health, and disease.* New York: Pantheon Books, 2013.

Patterson, N. et al., Genetic evidence for complex speciation of humans and chimpanzees. *Nature* 2006, ***441***(7097): 1103–8.

Schrenk, F., *Die Frühzeit des Menschen: Der Weg zum Homo sapiens.* Munich: Beck Verlag, 2019.

Lordkipanidze, D. et al., A complete skull from Dmanisi, Georgia, and the evolutionary biology of early *Homo*. *Science* 2013, ***342***(6156): 326–31.

Hublin, J. J. et al., New fossils from Jebel Irhoud, Morocco and the pan-African origin of *Homo sapiens*. *Nature* 2017, ***546***(7657): 289–92.

Berger, L. R. et al., *Homo naledi*, a new species of the genus *Homo* from the Dinaledi Chamber, South Africa. *eLife* 2015, ***4***.

Dirks, P. H. et al., Geological and taphonomic context for the new hominin species *Homo naledi* from the Dinaledi Chamber, South Africa. *eLife* 2015, ***4***.

Dirks, P. H. et al., The age of *Homo naledi* and associated sediments in the Rising Star Cave, South Africa. *eLife* 2017, ***6***: e24231.

Grun, R. et al., Dating the skull from Broken Hill, Zambia, and its position in human evolution. *Nature* 2020, ***580***(7803): 372–5.

Grun, R. et al., Direct dating of Florisbad hominid. *Nature* 1996, ***382***(6591): 500–1.

Kaiser, T. et al., Klimawandel als Antrieb der menschlichen Evolution. In H. Meller and T. Puttkammer, eds, *Klimagewalten: Treibende Kraft der Evolution.* Darmstadt: Konrad Theiss Verlag, 2017.

Chapter 4

Carbonell, E. et al., The first hominin of Europe. *Nature* 2008, ***452***(7186): 465–9.

Matsu'ura, S. et al., Age control of the first appearance datum for Javanese *Homo erectus* in the Sangiran area. *Science* 2020, ***367***(6474): 210–4.

Rizal, Y. et al., Last appearance of *Homo erectus* at Ngandong, Java, 117,000–108,000 years ago. *Nature* 2020, ***577***(7790): 381–5.

Lieberman, D. E., B. M. McBratney, and G. Krovitz, The evolution and development of cranial form in *Homo sapiens*. *PNAS* 2002, ***99***(3): 1134–9.

Stringer, C., The origin and evolution of *Homo sapiens*. *Philosophical Transactions of the Royal Society B: Biological Sciences* 2016, ***371***(1698). https://doi.org/10.1098/rstb.2015.0237.

Liu, W. et al., The earliest unequivocally modern humans in southern China. *Nature* 2015, ***526***(7575): 696–9.

Qiu, J., How China is rewriting the book on human origins. *Nature* 2016, ***535***: 22–5.

Fu, Q. et al., DNA analysis of an early modern human from Tianyuan Cave, China. *PNAS* 2013, ***110***(6): 2223–7.

Osipov, S. et al., The Toba supervolcano eruption caused severe tropical stratospheric ozone depletion. *Communications Earth & Environment* 2021, ***2***(1). doi: 10.1038/s43247-021-00141-7.

Krause, J. et al., The complete mitochondrial DNA genome of an unknown hominin from southern Siberia. *Nature* 2010, ***464***(7290): 894–7.

Zhang, D. et al., Denisovan DNA in Late Pleistocene sediments from Baishiya Karst Cave on the Tibetan Plateau. *Science* 2020, ***370***(6516): 584–7.

Sutikna, T. et al., Revised stratigraphy and chronology for *Homo floresiensis* at Liang Bua in Indonesia. *Nature* 2016, ***532***(7599): 366–9.

Morwood, M. J. et al., Archaeology and age of a new hominin from Flores in eastern Indonesia. *Nature* 2004, ***431***(7012): 1087–91.

Rampino, M. R., and S. Self, Bottleneck in human evolution and the Toba eruption. *Science* 1993, ***262***(5142): 1955.

Yu, H. et al., Palaeogenomic analysis of black rat (*Rattus rattus*) reveals multiple European introductions associated with human economic history. bioRxiv, 2021. https://www.biorxiv.org/content/10.1101/2021.04.14.439553v1.

Posth, C. et al., Pleistocene mitochondrial genomes suggest a single major dispersal of non-Africans and a Late Glacial population turnover in Europe. *Current Biology* 2016, ***26***(6): 827–33.

Beyer, R. M. et al., Windows out of Africa: A 300,000-year chronology of climatically plausible human contact with Eurasia. bioRxiv 2020. https://www.biorxiv.org/content/10.1101/2020.01.12.901694v1.

Beyer, R. M., M. Krapp, and A. Manica, High-resolution terrestrial climate, bioclimate and vegetation for the last 120,000 years. *Scientific Data* 2020, ***7***(1): 236.

Henshilwood, C. S. et al., Emergence of modern human behavior: Middle Stone Age engravings from South Africa. *Science* 2002, ***295***(5558): 1278–80.

Henshilwood, C. S. et al., A 100,000-year-old ochre-processing workshop at Blombos Cave, South Africa. *Science* 2011, ***334***(6053): 219–22.

Wadley, L. et al., Middle Stone Age bedding construction and settlement patterns at Sibudu, South Africa. *Science* 2011, ***334***(6061): 1388–91.

Lombard, M. and L. Phillipson, Indications of bow and stone-tipped arrow use 64 000 years ago in KwaZulu-Natal, South Africa. *Antiquity* 2010, ***84***: 635–48.

Backwell, L. et al., The antiquity of bow-and-arrow technology: Evidence from Middle Stone Age layers at Sibudu Cave. *Antiquity* 2018, ***92***(362): 289–303.

Armitage, S. J. et al., The southern route 'out of Africa': Evidence for an early expansion of modern humans into Arabia. *Science* 2011, ***331***(6016): 453–6.

Groucutt, H. S. et al., *Homo sapiens* in Arabia by 85,000 years ago. *Nature Ecology & Evolution* 2018, ***2***(5): 800–9.

Fu, Q. et al., An early modern human from Romania with a recent Neandertal ancestor. *Nature* 2015, ***524***(7564): 216–19.

Fu, Q. et al., Genome sequence of a 45,000-year-old modern human from western Siberia. *Nature* 2014, ***514***(7523): 445–9.

Fu, Q. et al., The genetic history of Ice Age Europe. *Nature* 2016, ***534***(7606): 200–5.

Lazaridis, I. et al., Ancient human genomes suggest three ancestral populations for present-day Europeans. *Nature* 2014, ***513***(7518): 409–13.

Lazaridis, I. et al., Genomic insights into the origin of farming in the ancient Near East. *Nature* 2016, ***536***(7617): 419–24.

Vernot, B., and J. M. Akey, Resurrecting surviving Neandertal lineages from modern human genomes. *Science* 2014, ***343***(6174): 1017–21.

Robock, A. et al., Did the Toba volcanic eruption of ~74k BP produce widespread glaciation? *Journal of Geophysical Research* 2009, ***114***. https://doi.org/10.1029/2008JD011652.

Chapter 5

Briggs, A. W. et al., Targeted retrieval and analysis of five Neandertal mtDNA genomes. *Science* 2009, ***325***(5938): 318–21.

Green, R. E. et al., A complete Neandertal mitochondrial genome sequence determined by high-throughput sequencing. *Cell* 2008, ***134***(3): 416–26.

Krause, J. et al., Neanderthals in central Asia and Siberia. *Nature* 2007, ***449***(7164): 902–4.

Sankararaman, S. et al., The genomic landscape of Neanderthal ancestry in present-day humans. *Nature* 2014, ***507***(7492): 354–7.

Dannemann, M., K. Prufer, and J. Kelso, Functional implications of Neandertal introgression in modern humans. *Genome Biology* 2017, ***18***(1): 61.

Dannemann, M., and J. Kelso, The contribution of Neanderthals to phenotypic variation in modern humans. *American Journal of Human Genetics* 2017, ***101***(4): 578–89.

Dannemann, M., A. M. Andres, and J. Kelso, Introgression of Neandertal- and Denisovan-like haplotypes contributes to adaptive variation in human toll-like receptors. *American Journal of Human Genetics* 2016, ***98***(1): 22–33.

COVID-19 Host Genetics Initiative, Mapping the human genetic architecture of COVID-19. *Nature* 2021, ***600***: 472–7.

Zeberg, H. et al., A Neanderthal sodium channel increases pain sensitivity in present-day humans. *Current Biology* 2020, ***30***(17): 3465–9, e4.

Zeberg, H., J. Kelso, and S. Pääbo, The Neandertal progesterone receptor. *Molecular Biology and Evolution* 2020, ***37***(9): 2655–2660.

Zeberg, H., and S. Pääbo, The major genetic risk factor for severe COVID-19 is inherited from Neanderthals. *Nature* 2020, ***587***(7835): 610–12.

Zeberg, H., and S. Pääbo, A genomic region associated with protection against severe COVID-19 is inherited from Neandertals. *PNAS* 2021, ***118***(9). doi: 10.1073/pnas.2026309118.

Giaccio, B. et al., High-precision (14)C and (40)Ar/(39)Ar dating of the Campanian Ignimbrite (Y-5) reconciles the time-scales of climatic–cultural processes at 40 ka. *Scientific Reports* 2017, ***7***: 45940.

Marti, A. et al., Reconstructing the plinian and co-ignimbrite sources of large volcanic eruptions: A novel approach for the Campanian Ignimbrite. *Scientific Reports* 2016, ***6***: 21220.

Krause, J., and T. Trappe, *Die Reise unserer Gene: Eine Geschichte über uns und unsere Vorfahren*. Berlin: Propyläen Verlag, 2019.

Dinnis, R. et al., New data for the Early Upper Paleolithic of Kostenki (Russia). *Journal of Human Evolution* 2019, ***127***: 21–40.

Krause, J. et al., A complete mtDNA genome of an early modern human from Kostenki, Russia. *Current Biology* 2010, ***20***(3): 231–6.

Hajdinjak, M. et al., Initial Upper Palaeolithic humans in Europe had recent Neanderthal ancestry. *Nature* 2021, ***592***(7853): 253–7.

Fellows Yates, J. A. et al., Central European woolly mammoth population dynamics: Insights from Late Pleistocene mitochondrial genomes. *Scientific Reports* 2017, ***7***(1): 17714.

Lorenzen, E. D. et al., Species-specific responses of Late Quaternary megafauna to climate and humans. *Nature* 2011, ***479***(7373): 359–64.

van der Kaars, S. et al., Humans rather than climate the primary cause of Pleistocene megafaunal extinction in Australia. *Nature Communications* 2017, ***8***: 14142.

Allentoft, M. E. et al., Extinct New Zealand megafauna were not in decline before human colonization. *PNAS* 2014, ***111***(13): 4922–7.

Remmert, H., The evolution of man and the extinction of animals. *Naturwissenschaften* 1982, ***69***(11): 524–7.

Napierala, H., A. W. Kandel, and N. J. Conard, Small game and shifting subsistence patterns from the Upper Palaeolithic to the Natufian at Baaz Rockshelter, Syria. In Marjan Mashkour and Mark Beech, eds, *Archaeozoology of the Near East 9*, Oxford: Oxbow, 2017.

Zvelebil, M., ed., *Hunters in transition: Mesolithic societies of temperate Eurasia and their transition to farming*. Cambridge: Cambridge University Press, 1986.

Greenberg, J., *A feathered river across the sky: The passenger pigeon's flight to extinction*. New York: Bloomsbury Publishing, 2014.

Sherkow, J. S., and H. T. Greely, What if extinction is not forever? *Science* 2013, ***340***: 32–3.

Blockstein, D. E., We can't bring back the passenger pigeon: The ethics of deception around de-extinction. *Ethics, Policy & Environment* 2017, ***20***: 33–7.

Shapiro, B., *How to clone a mammoth: The science of de-extinction*. Princeton, NJ: Princeton University Press, 2015.

Sikora, M. et al., The population history of northeastern Siberia since the Pleistocene. *Nature* 2019, ***570***(7760): 182–8.

Ardelean, C. F. et al., Evidence of human occupation in Mexico around the Last Glacial Maximum. *Nature* 2020, ***584***(7819): 87–92.

Holen, S. R. et al., A 130,000-year-old archaeological site in southern California, USA. *Nature* 2017, ***544***(7651): 479–83.

Posth, C. et al., Reconstructing the deep population history of Central and South America. *Cell* 2018, ***175***(5): 1185–97, e22.

Waters, M., and T. Stafford, The first Americans: A review of the evidence for the Late-Pleistocene peopling of the Americas. In Kelly Graf, Caroline Ketron, and Michael Waters, eds, *Paleoamerican Odyssey*. College Station: Texas A&M University Press, 2014.

Dillehay, T. D., and M. B. Collins, Early cultural evidence from Monte Verde in Chile. *Nature* 1988, ***332***: 150–2.

Pedersen, M. W. et al., Postglacial viability and colonization in North America's ice-free corridor. *Nature* 2016, ***537***(7618): 45–9.

Yu, H. et al., Paleolithic to Bronze Age Siberians reveal connections with first Americans and across Eurasia. *Cell* 2020, ***181***(6): 1232–45, e20.

Wang, C. C. et al., Genomic insights into the formation of human populations in East Asia. *Nature* 2021, ***591***(7850): 413–9.

Slon, V. et al., The genome of the offspring of a Neanderthal mother and a Denisovan father. *Nature* 2018, ***561***(7721): 113–6.

Slon, V. et al., A fourth Denisovan individual. *Science Advances* 2017, ***3***(7): e1700186.

Chen, F. et al., A late Middle Pleistocene Denisovan mandible from the Tibetan Plateau. *Nature* 2019, ***569***(7756): 409–12.

Zhang, D. et al., Denisovan DNA in Late Pleistocene sediments from Baishiya Karst Cave on the Tibetan Plateau. *Science* 2020, ***370***(6516): 584–7.

Huerta-Sanchez, E. et al., Altitude adaptation in Tibetans caused by introgression of Denisovan-like DNA. *Nature* 2014, ***512***(7513): 194–7.

Jeong, C. et al., Long-term genetic stability and a high-altitude East Asian origin for the peoples of the high valleys of the Himalayan arc. *PNAS* 2016, ***113***(27): 7485–90.

Zhang, X. et al., The history and evolution of the Denisovan-EPAS1 haplotype in Tibetans. *PNAS* 2021, ***118***(22).

Reich, D. et al., Genetic history of an archaic hominin group from Denisova Cave in Siberia. *Nature* 2010, ***468***(7327): 1053–60.

Browning, S. R. et al., Analysis of human sequence data reveals two pulses of archaic Denisovan admixture. *Cell* 2018, ***173***(1): 53–61 e9.

Cooper, A., and C. B. Stringer, Paleontology: Did the Denisovans cross Wallace's Line? *Science* 2013, ***342***(6156): 321–3.

Krause, J., Ancient human migrations. In H. S. R. Neck, ed., *Migration.* Vienna: Böhlau, 2011.

Diamond, J., Ten thousand years of solitude. *Discover* 1993, ***14***(3): 48–57.

Clark, J., Fanny Cochrane Smith (1834–1905). In *Australian Dictionary of Biography*, vol. 11. Melbourne: Melbourne University Press, 1988.

Chapter 6

Pääbo, S. et al., Genetic analyses from ancient DNA. *Annual Review of Genetics* 2004, ***38***: 645–79.

Louys, J., and P. Roberts, Environmental drivers of megafauna and hominin extinction in Southeast Asia. *Nature* 2020, ***586***(7829): 402–6.

Morwood, M. J. et al., Archaeology and age of a new hominin from Flores in eastern Indonesia. *Nature* 2004, ***431***(7012): 1087–91.

Jungers, W. L. et al., The foot of *Homo floresiensis. Nature* 2009 ***459***(7243): 81–4.

Detroit, F. et al., A new species of *Homo* from the Late Pleistocene of the Philippines. *Nature* 2019, ***568***(7751): 181–6.

Falk, D. et al., Brain shape in human microcephalics and *Homo floresiensis. PNAS* 2007, ***104***(7): 2513–8.

Culotta, E., Discoverers charge damage to 'hobbit' specimens. *Science* 2005, ***307***(5717). doi: 10.1126/science.307.5717.1848a.

Brumm, A. et al., Hominins on Flores, Indonesia, by one million years ago. *Nature* 2010, ***464***(7289): 748–52.

Brumm, A. et al., Early stone technology on Flores and its implications for *Homo floresiensis. Nature* 2006, ***441***(7093): 624–8.

Shine, R., and R. Somaweera, Last lizard standing: The enigmatic persistence of the Komodo dragon. *Global Ecology and Conservation* 2019, ***18***, e00624.

Smith, C. C., and S. D. Fretwell, The optimal balance between size and number of offspring. *American Naturalist* 1974, ***108***: 499–506.

Sutikna, T. et al., The spatio-temporal distribution of archaeological and faunal finds at Liang Bua (Flores, Indonesia) in light of the revised chronology for *Homo floresiensis*. *Journal of Human Evolution* 2018, ***124***: 52–74.

Jacobs, G. S. et al., Multiple deeply divergent Denisovan ancestries in Papuans. *Cell* 2019, ***177***(4): 1010–21, e32.

Bowler, J. M. et al., New ages for human occupation and climatic change at Lake Mungo, Australia. *Nature* 2003, ***421***(6925): 837–40.

Laidlaw, R., Aboriginal society before European settlement. In T. Gurry, ed., *The European occupation*. Richmond: Heinemann Educational Australia, 1984.

Cane, S., *First Footprints: The epic story of the first Australians*. St Leonards, New South Wales: Allen & Unwin, 2013.

Kik, A. et al., Language and ethnobiological skills decline precipitously in Papua New Guinea, the world's most linguistically diverse nation. *PNAS* 2021, ***118***(22). https://doi.org/10.1073/pnas.21000961.

Liberski, P. P., Kuru: A journey back in time from Papua New Guinea to the Neandertals' extinction. *Pathogens* 2013, ***2***(3): 472–505.

Tobler, R. et al., Aboriginal mitogenomes reveal 50,000 years of regionalism in Australia. *Nature* 2017, ***544***(7649): 180–4.

Hardy, M. C., J. Cochrane, and R. E. Allavena, Venomous and poisonous Australian animals of veterinary importance: A rich source of novel therapeutics. *BioMed Research International* 2014, ***2014***: 671041.

Perri, A. R. et al., Dog domestication and the dual dispersal of people and dogs into the Americas. *PNAS* 2021, ***118***(6). https://doi.org/10.1073/pnas.2010083118.

Frantz, L. A. et al., Genomic and archaeological evidence suggest a dual origin of domestic dogs. *Science* 2016, ***352***(6290): 1228–31.

Larson, G. et al., Rethinking dog domestication by integrating genetics, archeology, and biogeography. *PNAS* 2012, ***109***(23): 8878–83.

Thalmann, O. et al., Complete mitochondrial genomes of ancient canids suggest a European origin of domestic dogs. *Science* 2013, ***342***(6160): 871–4.

Leroy, G. et al., Genetic diversity of dog breeds: Between breed diversity, breed assignation and conservation approaches. *Animal Genetics* 2009, ***40***(3): 333–43.

Bergstrom, A. et al., Origins and genetic legacy of prehistoric dogs. *Science* 2020, ***370***(6516): 557–564.

Loog, L. et al., Ancient DNA suggests modern wolves trace their origin to a Late Pleistocene expansion from Beringia. *Molecular Ecology* 2020, ***29***(9): 1596–1610.

Baca, M. et al., Retreat and extinction of the Late Pleistocene cave bear (*Ursus spelaeus sensu lato*). *Naturwissenschaften* 2016, ***103***(11–12): 92.

Dugatkin, L. A., and L. Trut, *Füchse zähmen*. Berlin: Springer, 2017.

Wade, N., Nice rats, nasty rats: Maybe it's all in the genes. *New York Times*, 25 July 2006.

Plyusnina, I. Z. et al., Cross-fostering effects on weight, exploratory activity, acoustic startle reflex and corticosterone stress response in Norway gray rats selected for elimination and for enhancement of aggressiveness towards human. *Behavior Genetics* 2009, ***39***(2): 202–12.

Heyne, H. O. et al., Genetic influences on brain gene expression in rats selected for tameness and aggression. *Genetics* 2014, ***198***(3): 1277–90.

Albert, F. W. et al., Genetic architecture of tameness in a rat model of animal domestication. *Genetics* 2009, ***182***(2): 541–54.

Kukekova, A. V. et al., Red fox genome assembly identifies genomic regions associated with tame and aggressive behaviours. *Nature Ecology & Evolution* 2018, ***2***(9): 1479–91.

Lahr, Mirazón et al., Reply. *Nature* 2016, ***539***(7630): E10–E11.

Okin, G. S., Environmental impacts of food consumption by dogs and cats. *PLoS One* 2017, ***12***(8): e0181301.

Chapter 7

Boulanger, M. T., and R. L. Lyman, Northeastern North American Pleistocene megafauna chronologically overlapped minimally with Paleoindians. *Quaternary Science Reviews* 2014, ***85***: 35–46.

Ni Leathlobhair, M. et al., The evolutionary history of dogs in the Americas. *Science* 2018, ***361***(6397): 81–5.

Weissbrod, L. et al., Origins of house mice in ecological niches created by settled hunter-gatherers in the Levant 15,000 y ago. *PNAS* 2017, ***114***(16): 4099–104.

Peters, J. et al., Göbekli Tepe: Agriculture and Domestication. In Claire Smith, ed., *Encyclopedia of Global Archaeology*. New York: Springer, 2014.

van de Loosdrecht, M. et al., Pleistocene North African genomes link Near Eastern and sub-Saharan African human populations. *Science* 2018, ***360***(6388): 548–52.

Hublin, J. J. et al., New fossils from Jebel Irhoud, Morocco and the pan-African origin of *Homo sapiens*. *Nature* 2017, ***546***(7657): 289–92.

Humphrey, L. T. et al., Earliest evidence for caries and exploitation of starchy plant foods in Pleistocene hunter-gatherers from Morocco. *PNAS* 2014, ***111***(3): 954–9.

Lazaridis, I. et al., Genomic insights into the origin of farming in the ancient Near East. *Nature* 2016, ***536***(7617): 419–24.

Lazaridis, I. et al., Ancient human genomes suggest three ancestral populations for present-day Europeans. *Nature* 2014, ***513***(7518): 409–13.

Diamond, J., *Arm und Reich: Die Schicksale menschlicher Gesellschaften*. Frankfurt: Fischer Taschenbuch, 2006.

Hofmanova, Z. et al., Early farmers from across Europe directly descended from Neolithic Aegeans. *PNAS* 2016, ***113***(25): 6886–91.

Olalde, I. et al., The Beaker phenomenon and the genomic transformation of northwest Europe. *Nature* 2018, ***555***(7695): 190–6.

Chaplin, G., and N. G. Jablonski, Vitamin D and the evolution of human depigmentation. *American Journal of Physical Anthropology* 2009, ***139***(4): 451–61.

Skoglund, P. et al., Reconstructing prehistoric African population structure. *Cell* 2017, ***171***(1): 59–71, e21.

Wang, C. C. et al., Ancient human genome-wide data from a 3000-year interval in the Caucasus corresponds with ecogeographic regions. *Nature Communications* 2019, ***10***(1): 590.

Brandt, G. et al., Ancient DNA reveals key stages in the formation of central European mitochondrial genetic diversity. *Science* 2013, ***342***(6155): 257–61.

Mittnik, A. et al., The genetic prehistory of the Baltic Sea region. *Nature Communications* 2018, ***9***(1): 442.

Haak, W. et al., Massive migration from the steppe was a source for Indo-European languages in Europe. *Nature* 2015, ***522***(7555): 207–11.

Narasimhan, V. M. et al., The formation of human populations in South and Central Asia. *Science* 2019, ***365***(6457): eaat7487.

Shinde, V. et al., An ancient Harappan genome lacks ancestry from steppe pastoralists or Iranian farmers. *Cell* 2019, ***179***(3): 729–35, e10.

Oelze, V. M. et al., Early Neolithic diet and animal husbandry: stable isotope evidence from three Linearbandkeramik (LBK) sites in Central Germany. *Journal of Archaeological Science* 2010, ***38***: 270–9.

Bickle, P., and A. Whittle, *The first farmers of Central Europe: Diversity in LBK lifeways*. Oxford: Oxbow Books, 2013.

Schuenemann, V. J. et al., Ancient Egyptian mummy genomes suggest an increase of sub-Saharan African ancestry in post-Roman periods. *Nature Communications* 2017, ***8***: 15694.

Prendergast, M. E. et al., Ancient DNA reveals a multistep spread of the first herders into sub-Saharan Africa. *Science* 2019, ***365***(6448): eaaw6275.

Wang, K. et al., Ancient genomes reveal complex patterns of population movement, interaction, and replacement in sub-Saharan Africa. *Science Advances* 2020, ***6***(24): eaaz0183.

D'Andrea, A. C., T'ef (Eragrostis tef) in ancient agricultural systems of highland Ethiopia. *Economic Botany* 2008, ***62***(4): 547–66.

Schlebusch, C. M. et al., Southern African ancient genomes estimate modern human divergence to 350,000 to 260,000 years ago. *Science* 2017, ***358***(6363): 652–5.

de Filippo, C. et al., Bringing together linguistic and genetic evidence to test the Bantu expansion. *Proceedings of the Royal Society B: Biological Sciences* 2012, ***279***(1741): 3256–63.

Russell, T., F. Silva, and J. Steele, Modelling the spread of farming in the Bantu-speaking regions of Africa: an archaeology-based phylogeography. *PLoS One* 2014, ***9***(1): e87854.

Bostoen, K. et al., Middle to Late Holocene paleoclimatic change and the early Bantu expansion in the rain forests of western Central Africa. *Current Anthropology* 2015, ***56***: 354–84.

Tishkoff, S. A. et al., The genetic structure and history of Africans and African Americans. *Science* 2009, ***324***(5930): 1035–44.

Maxmen, A., Rare genetic sequences illuminate early humans' history in Africa. *Nature* 2018, ***563***(7729): 13–4.

Remmert, H., The evolution of man and the extinction of animals. *Naturwissenschaften* 1982, ***69***(11): 524–7.

Ning, C. et al., Ancient genomes from northern China suggest links between subsistence changes and human migration. *Nature Communications* 2020, ***11***(1): 2700.

Zhang, X. et al., The history and evolution of the Denisovan-EPAS1 haplotype in Tibetans. *PNAS* 2021, ***118***(22). doi: 10.1073/pnas.2020803118.

Xiang, H. et al., Origin and dispersal of early domestic pigs in northern China. *Scientific Reports* 2017, ***7***(1): 5602.

Hata, A. et al., Origin and evolutionary history of domestic chickens inferred from a large population study of Thai red junglefowl and indigenous chickens. *Scientific Reports* 2021, ***11***(1): 2035.

Posth, C. et al., Reconstructing the deep population history of Central and South America. *Cell* 2018, ***175***(5): 1185–97, e22.

Nakatsuka, N. et al., A paleogenomic reconstruction of the deep population history of the Andes. *Cell* 2020, ***181***(5): 1131–45, e21.

Swarts, K. et al., Genomic estimation of complex traits reveals ancient maize adaptation to temperate North America. *Science* 2017, ***357***(6350): 512–5.

Gutaker, R. M. et al., The origins and adaptation of European potatoes reconstructed from historical genomes. *Nature Ecology & Evolution* 2019, ***3***(7): 1093–101.

Shaw, B. et al., Emergence of a Neolithic in highland New Guinea by 5000 to 4000 years ago. *Science Advances* 2020, ***6***(13): eaay4573.

Bergstrom, A. et al., A Neolithic expansion, but strong genetic structure, in the independent history of New Guinea. *Science* 2017, ***357***(6356): 1160–3.

Carrington, D., Humans just 0.01 % of all life but have destroyed 83 % of wild mammals: Study. *Guardian*, May 21, 2018.

Chapter 8

Lipson, M. et al., Population turnover in remote Oceania shortly after initial settlement. *Current Biology* 2018, ***28***(7): 1157–65 e7.

McColl, H. et al., The prehistoric peopling of Southeast Asia. *Science* 2018, ***361***(6397): 88–92.

Habu, J., *Ancient Jomon of Japan*. Cambridge: Cambridge University Press, 2004.

Gakuhari, T. et al., Ancient Jomon genome sequence analysis sheds light on migration patterns of early East Asian populations. *Communications Biology* 2020, ***3***(1): 437.

Jinam, T. A. et al., Unique characteristics of the Ainu population in Northern Japan. *Journal of Human Genetics* 2015, ***60***(10): 565–71.

Posth, C. et al., Language continuity despite population replacement in Remote Oceania. *Nature Ecology & Evolution* 2018, ***2***(4): 731–40.

Skoglund, P. et al., Genomic insights into the peopling of the Southwest Pacific. *Nature* 2016, ***538***(7626): 510–3.

Skoglund, P., and D. Reich, A genomic view of the peopling of the Americas. *Current Opinion in Genetics and Development* 2016, ***41***: 27–35.

Wang, C. C. et al., Genomic insights into the formation of human populations in East Asia. *Nature* 2021, ***591***(7850): 413–9.

Gray, R. D., A. J. Drummond, and S. J. Greenhill, Language phylogenies reveal expansion pulses and pauses in Pacific settlement. *Science* 2009, ***323***(5913): 479–83.

Pugach, I. et al., Ancient DNA from Guam and the peopling of the Pacific. *PNAS* 2021, ***118***(1). https://doi.org/10.1073/pnas.2022112118.

Bellwood, P., *Man's conquest of the Pacific: The prehistory of Southeast Asia and Oceania*. Oxford: Oxford University Press, 1979.

Bellwood, P., *First migrants: Ancient migration in global perspective*. Oxford: Wiley Blackwell, 2014.

Kayser, M. et al., The impact of the Austronesian expansion: Evidence from mtDNA and Y chromosome diversity in the Admiralty Islands of Melanesia. *Molecular Biology and Evolution* 2008, ***25***(7): 1362–74.

Commendador, A. S. et al., A stable isotope (delta13C and delta15N) perspective on human diet on Rapa Nui (Easter Island) ca. AD 1400–1900. *American Journal of Physical Anthropology* 2013, ***152***(2): 173–85.

Clement, C. R. et al., Coconuts in the Americas. *Botanical Review* 2013, ***79***: 342–70.

Neel, J. V., Diabetes mellitus a 'thrifty' genotype rendered detrimental by 'progress'? *American Journal of Human Genetics* 1962, ***14***: 352–3.

O'Rourke, R. W., Metabolic thrift and the genetic basis of human obesity. *Annals of Surgery* 2014, ***259***(4): 642–8.

Lipo, C. P., T. L. Hunt, and S. Rapu Haoa, The "walking" megalithic statues (*moai*) of Easter Island. *Journal of Archaeological Science* 2013, ***40***(6). doi: 10.1016/j.jas.2012.09.029.

Diamond, J., *Collapse: How societies choose to fail or succeed*. New York: Viking, 2005.

Weisler, M. I., Centrality and the collapse of long-distance voyaging in east Polynesia. In M. D. Glascock, ed., *Geochemical evidence for long-distance exchange*. Westport, CT: Bergin and Garvey, 2002.

Kelly, L. G., Cook Island Origin of the Maori. *Journal of the Polynesian Society* 1955, ***64***(2): 181–96.

Knapp, M. et al., Mitogenomic evidence of close relationships between New Zealand's extinct giant raptors and small-sized Australian sister-taxa. *Molecular Phylogenetics and Evolution* 2019, ***134***: 122–8.

Valente, L., R. S. Etienne, and R. J. Garcia, Deep macroevolutionary impact of humans on New Zealand's unique avifauna. *Current Biology* 2019, ***29***(15): 2563–9, e4.

Giraldez, A., *The age of trade: The Manila galleons and the dawn of the global economy (exploring world history)*. Lanham, MD: Rowman & Littlefield, 2015.

Munoz-Rodriguez, P. et al., Reconciling conflicting phylogenies in the origin of sweet potato and dispersal to Polynesia. *Current Biology* 2018, ***28***(8): 1246–56 e12.

Borrell, B., DNA reveals how the chicken crossed the sea. *Nature* 2007, ***447***: 620–1.

Ioannidis, A. G. et al., Native American gene flow into Polynesia predating Easter Island settlement. *Nature* 2020, ***583***(7817): 572–7.

Pierron, D. et al., Genomic landscape of human diversity across Madagascar. *PNAS* 2017, ***114***(32): E6498-E6506.

Fernandes, D. M. et al., A genetic history of the pre-contact Caribbean. *Nature* 2021, ***590***(7844): 103–10.

Nagele, K. et al., Genomic insights into the early peopling of the Caribbean. *Science* 2020, ***369***(6502): 456–60.

Marcheco-Teruel, B. et al., Cuba: exploring the history of admixture and the genetic basis of pigmentation using autosomal and uniparental markers. *PLoS Genetics* 2014, ***10***(7): e1004488.

Flegontov, P. et al., Palaeo-Eskimo genetic ancestry and the peopling of Chukotka and North America. *Nature* 2019, ***570***(7760): 236–40.

Rasmussen, M. et al., Ancient human genome sequence of an extinct Palaeo-Eskimo. *Nature* 2010, ***463***(7282): 757–62.

McCartney, A. P., and J. M. Savelle, Thule Eskimo whaling in the Central Canadian Arctic. *Arctic Anthropology* 1985, ***22***(2): 37–58.

NCD Risk Factor Collaboration, Worldwide trends in body-mass index, underweight, overweight, and obesity from 1975 to 2016: A pooled analysis of 2,416 population-based measurement studies in 128.9 million children, adolescents, and adults. *Lancet* 2017, ***390***(10113): 2627–42.

Stevenson, C. M. et al., Variation in Rapa Nui (Easter Island) land use indicates production and population peaks prior to European contact. *PNAS* 2015, ***112***(4): 1025–30.

Wilmshurst, J. M. et al., Dating the late prehistoric dispersal of Polynesians to New Zealand using the commensal Pacific rat. *PNAS* 2008, ***105***(22): 7676–80.

Chapter 9

Radivojevic, M. et al., Painted ores and the rise of tin bronzes in Eurasia, c. 6500 years ago. *Antiquity* 2013, ***87***: 1030–45.

Anthony, D., *Horse, the wheel, and language: How Bronze-Age riders from the Eurasian Steppes shaped the modern world.* Princeton, NJ: Princeton University Press, 2010.

Levathes, L., *When China ruled the seas: The treasure fleet of the Dragon Throne, 1405–1433*, rev. edn. Oxford: Oxford University Press, 1997.

Wikipedia. History of the Great Wall of China. April 22, 2021. https://en.wikipedia.org/w/index.php?title=History_of_the_Great_Wall_of_China&oldid=1019348271.

Feng, Q. et al., Genetic history of Xinjiang's Uyghurs suggests Bronze Age multiple-way contacts in Eurasia. *Molecular Biology and Evolution* 2017, ***34***(10): 2572–82.

Ning, C. et al., Ancient genomes reveal Yamnaya-related ancestry and a potential source of Indo-European speakers in Iron Age Tianshan. *Current Biology* 2019, ***29***(15): 2526–32, e4.

Mallory, J. P., *The Tarim mummies: Ancient China and the mystery of the earliest peoples from the west.* London: Thames & Hudson, 2008.

Zhang, F. et al., The genomic origins of the Bronze Age Tarim Basin mummies. *Nature* 2021, ***599***: 256–61.

Allentoft, M. E. et al., Population genomics of Bronze Age Eurasia. *Nature* 2015, ***522***(7555): 167–72.

Gaunitz, C. et al., Ancient genomes revisit the ancestry of domestic and Przewalski's horses. *Science* 2018, ***360***(6384): 111–14.

Haak, W. et al., Massive migration from the steppe was a source for Indo-European languages in Europe. *Nature* 2015, ***522***(7555): 207–11.

Goldberg, A. et al., Ancient X chromosomes reveal contrasting sex bias in Neolithic and Bronze Age Eurasian migrations. *PNAS* 2017, ***114***(10): 2657–62.

Olalde, I. et al., The Beaker phenomenon and the genomic transformation of northwest Europe. *Nature* 2018, ***555***(7695): 190–6.

Olalde, I. et al., The genomic history of the Iberian Peninsula over the past 8,000 years. *Science* 2019, ***363***(6432): 1230.

Sommer, U., The appropriation or the destruction of memory? Bell Beaker "re-use" of older sites. In R. Bernbeck, K. P. Hofmann, and U. Sommer, eds., *Between memory sites and memory networks: New archaeological and historical perspectives.* Berlin: Edition Topoi, 2017.

Jeong, C. et al., A dynamic 6,000-year genetic history of Eurasia's Eastern Steppe. *Cell* 2020, ***183***(4): 890–904, e29.

Taylor, W. T. T. et al., Evidence for early dispersal of domestic sheep into Central Asia. *Nature Human Behaviour* 2021, ***5***: 1169–79.

Krzewinska, M. et al., Ancient genomes suggest the eastern Pontic-Caspian Steppe as the source of western Iron Age nomads. *Science Advances* 2018, ***4***(10): eaat4457.

Ventresca Miller, A. et al., Subsistence and social change in central Eurasia: Stable isotope analysis of populations spanning the Bronze Age transition. *Journal of Archaeological Science* 2013, ***42***: 525–38.

Lindner, S., Chariots in the Eurasian Steppe: A Bayesian approach to the emergence of horse-drawn transport in the early second millennium BC. *Antiquity* 2020, ***94***(374).

Mittnik, A. et al., Kinship-based social inequality in Bronze Age Europe. *Science* 2019, ***366***(6466): 731–4.

Narasimhan, V. M. et al., The formation of human populations in South and Central Asia. *Science* 2019, ***365***(6457): eaat7487.

Reich, D. et al., Reconstructing Indian population history. *Nature* 2009, ***461***(7263): 489–94.

Andrades Valtuena, A. et al., The Stone Age plague and its persistence in Eurasia. *Current Biology* 2017, ***27***(23): 3683–91, e8.

Rasmussen, S. et al., Early divergent strains of *Yersinia pestis* in Eurasia 5,000 years ago. *Cell* 2015, ***163***(3): 571–82.

Rascovan, N. et al., Emergence and spread of basal lineages of *Yersinia pestis* during the Neolithic decline. *Cell* 2019, ***176***(1–2): 295–305, e10.

Hinnebusch, B. J., The evolution of flea-borne transmission in *Yersinia pestis*. *Current Issues in Molecular Biology* 2005, ***7***(2): 197–212.

Spyrou, M. A. et al., Analysis of 3800-year-old *Yersinia pestis* genomes suggests Bronze Age origin for bubonic plague. *Nature Communications* 2018, ***9***(1): 2234.

Spyrou, M. A. et al., Ancient pathogen genomics as an emerging tool for infectious disease research. *Nature Reviews Genetics* 2019, ***20***(6): 323–40.

Feldman, M. et al., Ancient DNA sheds light on the genetic origins of early Iron Age Philistines. *Science Advances* 2019, ***5***(7): eaax0061.

Trevisanato, S. I., The "Hittite plague," an epidemic of tularemia and the first record of biological warfare. *Medical Hypotheses* 2007, ***69***(6): 1371–4.

Gnecchi-Ruscone, G. A. et al., Ancient genomic time transect from the Central Asian Steppe unravels the history of the Scythians. *Science Advances* 2021, ***7***(13). doi: 10.1126/sciadv.abe4414.

Neparaczki, E. et al., Y-chromosome haplogroups from Hun, Avar and conquering Hungarian period nomadic people of the Carpathian Basin. *Scientific Reports* 2019, ***9***(1): 16569.

Nagy, P. L. et al., Determination of the phylogenetic origins of the Arpad Dynasty based on Y chromosome sequencing of Bela the Third. *European Journal of Human Genetics* 2021, ***29***(1): 164–72.

Weatherford, J., *Genghis Khan and the making of the modern world.* Grand Haven, MI: Brilliance Publishing, 2014.

Wheelis, M., Biological warfare at the 1346 siege of Caffa. *Emerging Infectious Diseases* 2002, ***8***(9): 971–5.

Schmid, B. V. et al., Climate-driven introduction of the Black Death and successive plague reintroductions into Europe. *PNAS* 2015, ***112***(10): 3020– 5.

Deaton, A., *Der große Ausbruch: Von Armut und Wohlstand der Nationen.* Stuttgart: Klett-Cotta, 2017.

Chapter 10

Benedictow, O., *Black Death 1346–1353: The Complete History*. Suffolk: Boydell Press, 2006.

Gottfried, R. S., *Black Death: Natural and Human Disaster in Medieval Europe*. New York: Free Press, 1985.

Mann, M. E. et al., Global signatures and dynamical origins of the Little Ice Age and Medieval Climate Anomaly. *Science* 2009, ***326***(5957): 1256–60.

Kintisch, E., Why did Greenland's Vikings disappear? *Science*, 2016, November 10. https://www.science.org/content/article/why-did-greenland-s-vikings-disappear.

Ladurie, E. L. R., *Times of feast, times of famine: A history of climate since the year 1000*. New York: Allen & Unwin, 1988.

Ruddiman, W. F., *Earth's climate past and future*, 3rd edn. New York: W. H. Freeman, 2013.

Stocker, B. D. et al., Holocene peatland and ice-core data constraints on the timing and magnitude of CO_2 emissions from past land use. *PNAS* 2017, ***114***(7): 1492–7.

Keller, M. et al., Ancient *Yersinia pestis* genomes from across Western Europe reveal early diversification during the First Pandemic (541–750). *PNAS* 2019, ***116***(25): 12363–72.

Stathakopoulos, D. C., *Famine and Pestilence in the Late Roman and Early Byzantine Empire: A Systematic Survey of Subsistence Crises and Epidemics*. London: Taylor & Francis, 2004.

Shaw-Taylor, L., An introduction to the history of infectious diseases, epidemics and the early phases of the long-run decline in mortality. *Economic History Review* 2020, ***73***(3): E1–E19.

Kocher, A. et al., Ten millennia of hepatitis B virus evolution. *Science* 2021, ***374***(6564): 182–8.

Vagene, A. J. et al., *Salmonella enterica* genomes from victims of a major sixteenth-century epidemic in Mexico. *Nature Ecology & Evolution* 2018, ***2***(3): 520–8.

Barquera, R. et al., Origin and health status of first-generation Africans from early colonial Mexico. *Current Biology* 2020, ***30***(11): 2078–91, e11.

Kaner, J., and S. Schaack, Understanding Ebola: The 2014 epidemic. *Global Health* 2016, ***12***(1): 53.

Chaib, F., *New report calls for urgent action to avert antimicrobial resistance crisis*. New York: World Health Organization, 2019.

Taubenberger, J. K. et al., Characterization of the 1918 influenza virus polymerase genes. *Nature* 2005, ***437***(7060): 889–93.

Taubenberger, J. K., and D. M. Morens, Influenza: the once and future pandemic. *Public Health Reports* 2010, ***125***(Suppl 3): 16–26.

Cordaux, R., and M. A. Batzer, The impact of retrotransposons on human genome evolution. *Nature Reviews Genetics* 2009, ***10***(10): 691–703.

Zhang, G. et al., Comparative genomics reveals insights into avian genome evolution and adaptation. *Science* 2014, ***346***(6215): 1311–20.

Hormozdiari, F. et al., Rates and patterns of great ape retrotransposition. *PNAS* 2013, ***110***(33): 13457–62.

Prufer, K. et al., The bonobo genome compared with the chimpanzee and human genomes. *Nature* 2012, ***486***(7404): 527–31.

Gunz, P. et al., Neandertal introgression sheds light on modern human endocranial globularity. *Current Biology* 2019, ***29***(5): 895.

Neubauer, S., J. J. Hublin, and P. Gunz, The evolution of modern human brain shape. *Science Advances* 2018, ***4***(1): eaao5961.

Harari, Y. N., *21 Lessons for the 21st Century*. London: Jonathan Cape, 2018.

Wesson, P., Cosmology, extraterrestrial intelligence, and a resolution of the Fermi–Hart paradox. *Royal Astronomical Society* 1992, ***31***: 161–70.

Annex sources

Takahashi, K., and S. Yamanaka, Induction of pluripotent stem cells from mouse embryonic and adult fibroblast cultures by defined factors. *Cell* 2006, ***126***(4): 663–76.

Jinek M. et al., A programmable dual-RNA-guided DNA endonuclease in adaptive bacterial immunity. *Science* 2012, ***337***(6096): 816–21.

Hublin, J. J., Out of Africa: Modern human origins special feature: the origin of Neandertals. *PNAS* 2009, ***106***(38): 16022–7.

Stringer, C., and P. Andrews, *The complete world of human evolution*, rev. edn. London; New York: Thames & Hudson, 2011.

Sudlow, C. et al., UK biobank: An open access resource for identifying the causes of a wide range of complex diseases of middle and old age. *PLoS Medicine* 2015, ***12***(3): e1001779.

Ma, R. et al., Hepatitis B virus infection and replication in human bone marrow mesenchymal stem cells. *Virology Journal* 2011, ***8***: 486.

List of Figures

Acknowledgements

We would like to thank Wolfgang Haak, Alexander Herbig, Kathrin Nägele, Svante Pääbo, Walter Pohl, Kay Prüfer, Harald Ringbauer, Stephan Schiffels and Philipp Stockhammer for proofreading individual chapters.

The insights into human evolution and the genetic history of human populations discussed in this book would not have been possible without the research of countless academics from the fields of archaeology, anthropology, bioinformatics, genetics, history, linguistics, and medicine. Without them we would never have been able to reconstruct numerous episodes from human history. For this, special thanks go to Kurt Alt, Luca Bandioli, Natalia Berezina, Hervé Bocherens, Madelaine Böhme, Abdeljalil Bouzouggar, Adam Brumm, Jaroslav Bružek, Jane Buikstra, Hernan Burbano, Alexandra Buzhilova, David Caramelli, Nicholas Conard, Alfredo Coppa, Isabelle Crevecoeur, Yinqiu Cui, Yadira de Armas, Anatoli Derevjanko, Leyla Djansugurova, Dorothée Drucker, Qiaomei Fu, Patrick Geary, Richard Edward Green, Mateja Hajdinjak, Svend Hansen, Michaela Harbeck, Katerina Harvati, Jean-Jacques Hublin, Daniel Huson, Janet Kelso, Martin Kircher, Egor Kitov, Corina Knipper, Kristian Kristiansen, Carles Lalueza Fox, Greger Larson, Iosif Lazaridis, Mark Lipson, Sandra Lösch, Anna-Sapfo Malaspinas, Tomislav Maricic, Iain Mathieson, Michael McCormick, Harald Meller, Matthias Meyer, Christopher Miller, Kay Nieselt, Inigo Olalde, Ludovic Orlando, Svante Pääbo, Nick Patterson, Ernst Pernicka, Benjamin Petter, Ron Pinhasi, David Reich, Sabine Reinhold, Roberto Risch, Patrick Roberts, Mirjana Roksandic, Hélèn Rougier, Eleanor Scerri, Hannes Schröder, Patrick Semal, Pontus Skoglund, Montgomery Slatkin, Viviane Slon, Anne Stone, Mark Stoneking, Jiri Svoboda, Anna Szecsenyi-Nagy, Frédérique Valentin, Petr Veleminsky, Benjamin Vernot, Tivadar Vida, Joachim Wahl, Hugo Zeberg – and many other unnamed but highly esteemed colleagues.

We would also like to thank the staff and colleagues at the University of Tübingen as well as the Max Planck Institute for Human History in Jena and the Max Planck Institute for Evolutionary Anthropology in Leipzig. Special thanks go to Rodrigo Barquera, Kirsten Bos, Selina Carlhoff, Michal Feldman, Russel Gray, Wolfgang Haak, Alexander Herbig, Zuzana Hofmanová, Choongwon Jeong, Marcel Keller, Felix Key, Arthur Kocher, Denise Kühnert, Thiseas Lamnidis, Angela Mötsch, Lyazzat Musralina, Kathrin Nägele, Cosimo Posth, Adam Powell, Harald Ringbauer, Guido Gnecchi Ruscone, Stephan Schiffels, Verena Schünemann, Eirini Skourtanioti, Maria Spyrou, Philipp Stockhammer, Rezeda Tukhbatova, Marieke van der Loosdrecht, Åshild Vågene, Vanessa Villalba, Chuanchao Wang, Ke Wang, Christina Warinner, He Yu, and all the others who were instrumental to the projects described in this book.

At Propyläen Verlag we were assisted by Kristin Rotter, particularly in the design of the book, and by Heike Wolter in the finer details. Our thanks to Tom Bjoerklund, Falko Daim, Elisabeth Daynes, John Gurche, Christopher Henshilwood, Rüdiger Krause, Walter Pohl, Peter Schouten, Velizar Simeonovski and Damian Wolny for the wonderful illustrations they contributed to this book. We thank Peter Palm for his clearly structured maps. And we thank our agent, Franziska Günther, for her assistance – not least with deciding on a title for the book.

Johannes Krause would like to thank his wife Henrike for the many discussions of the book and for her tolerance of long evenings spent in front of the computer. He would like to apologize to his daughter Alma Edda for the neglected bedtime stories. And he would also like to thank his parents, Maria and Klaus-Dieter, his sister Kristin, and Rebekka Büttner for reading and providing constructive comments on the manuscript as a whole. Thomas Trappe wishes to thank his entire family – especially Claudia, without whom it would have been impossible to juggle a challenging new job and this book in the exceptional circumstances of a pandemic that lasted for one and a half years. Particular thanks are due to Clara and Leo, the finest beings on this planet, for their strength in a year that weighed heavy on them. The future belongs to you – I know you will make the best of it.

Notes

Notes to Chapter 1

1 There are embryonic stem cells that can form whole organisms, but these are no longer used for the reproduction of cell tissue for ethical reasons. On the other hand, the use of so-called pluripotent stem cells is ethically and legally acceptable. Unlike totipotent embryonic stem cells, these cannot be used to grow whole organisms, but only parts of them – such as liver, heart or brain cells. As early as 2006, the Japanese researcher Shin'ya Yamanaka developed a technique for producing pluripotent stem cells from any body cell, through the induction of a few genes. This discovery won Yamanaka the Nobel Prize in Physiology or Medicine in 2012, not least because it made stem cell research possible without the need to use embryos.

2 The CRISPR/Cas9 method, better known as 'gene scissors', was developed in 2021 by a team headed by the French researcher Emmanuelle Charpentier and the American molecular biologist Jennifer Doudna. Both women were awarded the Nobel Prize for Chemistry in 2020. Their discovery ushered in a new age of molecular biological medicine and genetic research. Gene scissors exploit a mechanism that is present in about half of all known bacterial organisms. The CRISPR/Cas9 system allows bacteria to store in their cell structure some of the genetic information from the viruses that attack them and thus to build up an immune response that can then be passed on to the next generation. If at a later stage the same type of virus re-encounters the modified bacterial cell, it will activate its immune memory. Bacterial RNA molecules will then attack the basal combinations recognized by the bacteria from the previous infection. Finally, the cas enzyme released by the bacteria as a rearguard action sets to work, cutting out the piece of virus marked by the RNA. The virus is, as it were, dissected with the gene scissors and thus rendered harmless. The CRISPR/Cas9 technique, which was copied from the bacteria themselves, is now part of the standard toolkit of genetic research. In November 2019, the US–Swiss company CRISPR Therapeutics, of which Emmanuelle Charpentier is a co-founder, announced success in two patients with the inherited blood disorders beta thalassemia and sickle cell anaemia;

this success was soon followed by promising progress in the treatment of cancer patients. Both cases involved cutting out, from the DNA of the body's own immune cells, information that normally serves to inhibit the immune response.

3 Although we know the ninety genes that distinguish us from the Neandertals, it is not genes alone that are responsible for different phenotypes (the physical characteristics of cells). Rather it is the combination of different genetic variants, the number of modifications in the genes, and their interaction with other bases in the genome: all these factors determine which proteins and how many of them are produced in the various cells. This is also the reason why geneticists can easily decode the function of individual genes in single-cell organisms, yet are still none the wiser when it comes to humans.

4 However, this is not the same 2 per cent but a large pool of genetic information, from which each receives a small handful. Thus around 40 per cent of the genetic information of the Neandertal and more than half of the Denisovan's is still floating around in the great ocean that constitutes the human genome; the rest has disappeared without trace.

5 This task can be imagined as a giant Neandertal jigsaw, very similar to a modern human jigsaw. From the Neandertals' pool of genetic information, pieces are fished out that are identical in colour and shape with those of the master template, in other words with our genetic code. Bit by bit, a picture appears, but there are still many gaps. In the Neandertal puzzle we can find a piece with the right shape to fit the gap, but a completely different colour; thus one of the fixed differences is rendered visible. As mentioned, back then the Neandertal jigsaw was one with many missing pieces, and by the end of five years of work only about a half of the genome was in sight.

6 We are already able to transplant organs across different species. Doctors and researchers hope that such xenotransplantations will soon offer a solution to the undersupply of organs in human medicine.

7 Not every mutation results in changed inherited characteristics, far from it; only a very few do. This is because only 2 per cent of the genome actually contains genes – the rest serves to control the genes, and more than half has no known function. This part is also referred to as 'junk DNA'.

Note to Chapter 2

1 It should be added that the genetic information of this mysterious last common ancestor didn't come from a discovered bone – which would have been a huge archaeological sensation and an incredible fluke – but was calculated from the genome of a living modern human.

Notes to Chapter 3

1 This doesn't contradict the fact that the only fully sequenced genomes we have so far are from female Neandertals. The Y chromosome – which, unlike the female sex chromosome (XX), occurs only singly (XY) – has to undergo a complex procedure to be fished out from the relevant samples. It is coincidental that all the genomes we have managed to reconstruct in full are from Neandertal women.

2 According to the theory of multi-regionalism, which was still defended in the noughties, only Europeans are descended from the Neandertals, just as the origins of humans in Africa and Asia were thought to lie in primitive human types native to those continents. This theory has since been disproved, not least on the basis of the sequencing of the Neandertal genome: thanks to it, we now know that humans carry only a small proportion of Neandertal DNA, and only outside sub-Saharan Africa. In other words the genes confirm the out-of-Africa theory, according to which modern humans evolved on that continent and spread from there.

3 In keeping with Bergmann's rule, according to which living things are on average bulkier the colder the region they live in, Neandertals, too, had a climate-adjusted ratio of body mass to body surface, meaning that they lost less heat through their skin than their larger human relatives in the south.

4 *Homo rhodesiensis* is commonly referred to as the African *Homo heidelbergensis.*

Notes to Chapter 4

1 For such crossings to give rise to subsequent populations, logic dictates that at least one woman and one man must have made the journey – but probably far more in actual fact, otherwise the gene pool created through a few generations of incest would give migrants poor chances of long-term survival.

2 Ghost populations are commonly encountered in archaeogenetics, for example in the genomes of some contemporary Africans or Americans of African origin. Their DNA can contain components whose gene clock points to an admixture that sometimes dates back several hundreds of thousands of years: thus, as well as mixing with the Neandertals in the north, modern humans seem to have done the same in sub-Saharan Africa, but with an unknown early African hominin.

3 The term 'basal' refers to the very early split from the genetic lineage that Europeans and Asians were later to develop from.

4 For simplicity's sake, the term 'Natufian' is used here and subsequently to denote people associated with the Natufian culture, rather than a clearly definable population.

Note to Chapter 5

1 This value was calculated for the United Kingdom, but is relevant to the whole of Western Europe. The main reason why the United Kingdom is so often cited as a reference in medical genome research is that it has by the far the largest genome dataset of all European countries. Such data are virtually non-existent in Germany.

Notes to Chapter 6

1 For the avoidance of doubt, it should be added that the fictional character of Tarzan lived in the African jungle.

2 As long ago as 1999, the Boston-based evolutionary biologist Jared Diamond summed up, in his groundbreaking book *Guns, Germs, and Steel: The Fates of Human Societies*, which factors were key to the ability to turn wild animals into domestic and edible ones. (The dog thus falls into a special category of domestic animal.) According to Diamond, only herbivores are suitable for this purpose, as otherwise the protein output would be no greater than the protein input. Rapid growth and hence a high reproduction rate are also necessary in order to make the most efficient use of the animal products. And only those animals can be used that will reproduce in captivity and are accustomed to living in herds without generating status and territorial conflicts. Highly aggressive animals can be ruled out from the start, Diamond argues, citing the bear as an example. Taking all this into account, he writes, there was a much greater choice of domesticable candidates in Europe than in Africa.

Notes to Chapter 7

1 Just like 'Natufians', this term is used to designate people who were associated with the Iberomaurusian culture; here again, there is no clearly definable population.

2 The proportions vary considerably within Europe, however. Present-day Sards carry less than 5 per cent hunter-gatherer DNA, for example, while Estonians have more than 60 per cent.

3 The people associated with this culture entered Europe with brute force 4,900 years ago and caused the last massive upheaval in the continent's DNA. On average, today's Europeans owe around 75 per cent of their DNA to both populations of the Fertile Crescent, while the rest goes back to the hunter-gatherers of eastern and western Europe.

4 In those days rice was not grown in water, like now, but as a dry crop. Both methods work, but the modern method became established because it prevents the massive weed growth in tropical regions without impairing the growth of the rice plant itself.

5 Five thousand years ago, the human population had already grown to 14 million, from a figure probably in the single-digit millions at the beginning of the Holocene; by the beginning of the Christian era it was more than ten times that size.

Note to Chapter 8

1 It extended from the Bismarck Archipelago in the west to New Caledonia in the south, Samoa in the east, and Tuvalu in the north.

Notes to Chapter 10

1 Unlike bacteria, viruses mainly reproduce not in the blood but in the soft tissue of the human body – the lungs in influenza and coronaviruses, for example, and the liver in hepatitis B (HBV). Since the human bloodstream includes the bones and teeth, bacteria will be found here too if the analysed individual suffered a bacterial blood infection. Consequently, in the best-case scenario, the corpse of a plague victim will remain suffused with bacterial DNA for millennia. Soft tissue, by contrast, decays within weeks of death, along with any viruses it contains. That is, unless the organs are deep-frozen in the permafrost or the soft tissue is itself surrounded by bone – as in the case of HBV viruses, which infect the liver primarily but then retreat into the bone marrow after being attacked by the immune system. Sequencing is also aided by the structure of these viruses, which do not consist of RNA but of the much more stable DNA.

2 HBV must have been more or less endemic among hunter-gatherers: a persistent health problem, in other words. Numerous branches of the HBV strain developed, including in Neolithic Europe, after the transition from apes to humans. This diversity then came to an end here, 3,200 years ago, and a new form of hepatitis B did not appear until 400 years later. This gap coincides with the suspected outbreak of the Bronze Age plague: it is conceivable therefore that the pathogen lost its home as a result of mass deaths during this period, but survived in human populations outside the epicentre of the pandemic. The HBV strain that gave rise to the present-day pathogen came from a variant that resurfaced in Europe just over 2,800 years ago. This finding suggests a dramatic drop in population in the preceding millennia, much like during the Stone Age plague 4,800 years ago.

3 In HIV sufferers, the white blood cells incorporate the virus into their genome and go on to produce new viruses according to the infiltrated blueprint before they themselves are eventually destroyed. If too many white blood cells are destroyed, the immune deficiency disease AIDS develops. Although there is no

cure for this disease to date, it can be kept at bay very effectively with drugs designed to inhibit the activity of the retroviruses.

4 The Doomsday Clock was invented in 1947 by the journal *Bulletin of the Atomic Scientists*. It is designed to illustrate how close the world is to Armageddon.

Index

Page numbers in *italics* refer to figures.